STUDENT MATHEMATICAL LIBRARY
Volume 8

EXPLORING THE NUMBER JUNGLE:
A JOURNEY INTO DIOPHANTINE ANALYSIS

EDWARD B. BURGER

AMERICAN MATHEMATICAL SOCIETY

Opening Thoughts:
Welcome to the Jungle

One of the most fundamental notions in mathematics is that of number. Although the idea of number is basic, the numbers themselves possess both nuance and complexity that spark the imagination.

> *Mathematics is the queen of the sciences and number theory is the queen of mathematics.*
>
> — Carl Friedrich Gauss

Welcome to diophantine analysis—an area of number theory in which we attempt to discover hidden treasures and truths within the jungle of numbers by exploring the rational numbers. Diophantine analysis comprises two different but interconnected domains—diophantine approximation and diophantine equations.

The rational numbers are creatures that appear both familiar and understandable. In comparison, the irrational numbers, in some sense, seem completely enigmatic. If we think about the rational numbers as sitting in the real line, then diophantine approximation can be viewed as an exploration of the number line under a microscope. If we view a piece of the real number line under higher and higher magnification we would see the following:

Given this brief life history, your first challenge now awaits:

How old was Diophantus when he died?

Little background is required for the journey ahead, and a familiarity with number theory is not expected. An understanding of calculus and basic linear algebra together with the desire and ability to prove theorems are all that is needed for most of the material. Toward the end of this book, a few modules require some familiarity with beginning abstract algebra. Each mathematical theme is presented in a self-contained manner and is motived by very basic notions. However beware: There are a few major jolts along the way—some extremely challenging questions and even some open problems are thrown in!

In this book, you as the reader will be an active participant in the explorations that lie ahead. In fact, the cover art illustrates this theme—hidden in the front and back covers are "magic" stereogram images. The secret to seeing the hidden images is to gaze at the cover but focus (converge your eyes) on a position twice as far away as the cover. After a few minutes the 3-D image should appear (helpful viewing hints are included on the inside of the front cover). Thus by actively participating, you will be able to discover the otherwise invisible structure and richness in the number jungle.

Each module comprises a sequence of numbered questions that you are asked to answer and statements that you are asked to verify. Many hints and remarks are included and should be freely utilized and enjoyed. The hints do not reveal any answers; instead they are designed to give a gentle push in a particular direction. They are placed at the end of the book only to allow readers, if they wish, to think about the issues first without being distracted. There is no wrong way of using the hints, and readers are welcome to immediately visit them if they wish. There are three types of (star-struck) hints, and throughout the text you will see the symbols [⋆ HINT], [⋆⋆ HINT], and [⋆⋆⋆ HINT]. A one-starred hint provides a tiny nudge to help get you started. A two-starred hint contains more extensive remarks and suggestions, and a three-starred hint is more significant and perhaps includes an outline or overview of a plan of attack. Each module closes with a Big Picture Question that invites you to step back from all the technical details and take a panoramic view of how the ideas at hand fit into the larger mathematical landscape. Commentaries for selected Big Picture Questions are also included at the end of the book. With the text as your guide and your own creative contributions as the points of interest, I hope you find your journey to be a truly interactive experience.

> *...the primary question was not What we know,*
> *but How do we know it.*
>
> — Aristotle

Although this book is small, its aims are not. The book is an invitation to develop the ideas of diophantine analysis and to discover the fundamental results on your own. I believe that this subject is one of the most beautiful and rich areas of mathematics, and I hope you will share my point of view and enjoy the sights and challenges that lie ahead. This book also provides a setting where different areas and ideas of mathematics come together. Here you will see basic notions from algebra, analysis, geometry, and topology join together as one collective whole. Most importantly however, I hope that through the process of actively generating the ideas behind the subject, you will continue to hone the habits of thought required to analyze issues and reason in a more effective and creative manner.

Uncovering the mysterious structure of number has been one of the great intellectual challenges of human history. I hope you enjoy your journey through the beautiful mathematics that lies ahead and enjoy the challenge of conquering this area of number theory for yourself. *You can do it!*

With all good wishes,

Edward Burger
January 1, 2000

1.2. Give an example of a subset $X \subseteq \mathbb{R}$ that is neither dense nor discrete in $\mathbb{R}$. Justify your answer.

We write $\mathbb{Z}$ to denote the set of integers. (The symbol $\mathbb{Z}$ evolved from the German word *Zahlen* which means number.)

1.3. Is $\mathbb{Z}$ a dense or discrete subset of $\mathbb{R}$?

We now introduce the main players in our story: We write $\mathbb{Q}$ for the set of rational numbers. (It is an exercise for the etymologically inclined reader to figure the origins of the symbol $\mathbb{Q}$.)

1.4. Is $\mathbb{Q}$ a dense or discrete subset of $\mathbb{R}$?

Let $\mathfrak{Q}_N$ be the set of all rational numbers having denominators not exceeding N.

1.5. Verify that

$$\bigcup_{N=1}^{\infty} \mathfrak{Q}_N = \mathbb{Q}.$$

As a first attempt to understand how the rational numbers sit in the real number line, we consider the analogous, albeit easier, issue for the set $\mathfrak{Q}_N$.

1.6. How close can two distinct elements of $\mathfrak{Q}_N$ be? Conjecture a value for

$$\min\{|r_1 - r_2| : r_1, r_2 \in \mathfrak{Q}_N, r_1 \neq r_2\},$$

and then prove your conjecture.

1.7. Is $\mathfrak{Q}_N$ a dense subset of $\mathbb{R}$, a discrete subset of $\mathbb{R}$, or neither?

Here is our first diophantine approximation result which captures the spirit of the subject.

THEOREM 1.8. *Let $\alpha \in \mathbb{R}$ and N be a positive integer. Then there exists an element $r \in \mathfrak{Q}_N$ such that*

$$|\alpha - r| \leq \frac{1}{N}.$$

In fact two simple corollaries of the previous theorem foreshadow much of what is to come.

COROLLARY 1.9. *Let α be a real number and N a positive integer. Then there exists a rational number p/q such that $1 \leq q \leq N$ and*

$$\left| \alpha - \frac{p}{q} \right| \leq \frac{1}{N}.$$

COROLLARY 1.10. *Let $\alpha \in \mathbb{R}$ be an irrational number. Then there exist infinitely many rational numbers p/q that satisfy*

$$\left| \alpha - \frac{p}{q} \right| < \frac{1}{q}.$$

[⋆⋆ HINT]

We can use Corollary 1.9 to give an elegant, quantitative answer to Question 1.4.

1.11. Use Corollary 1.9 to prove that $\mathbb{Q}$ is a dense subset of $\mathbb{R}$.

Even though Corollary 1.9 is not a particularly deep result, it cannot be improved for certain choices of α. That is, given an integer N, there are α for which we would have *equality* in the upper bounds of the two inequalities of Corollary 1.9. Thus we say that the inequalities are *sharp* for those α's.

1.12. Given a fixed integer N, for what value(s) of α are the upper bounds of the inequalities in Corollary 1.9 sharp?

We've seen that Corollary 1.9 is sharp for some values of α. But this observation does not preclude the possibility that Corollaries 1.9 and 1.10 can be improved further. Our goal will be to find the best possible bounds for such inequalities. This goal and variations on the theme are the main focuses of basic diophantine approximation.

We've just shown that rational numbers can approximate arbitrary real numbers fairly well in terms of the size of the denominator of the rational approximates. Let's see how well we can approximate rational numbers by *other* rational numbers in terms of the sizes of their denominators. In this direction, let's find a *lower bound* for the distance between any two distinct rational numbers.

$$\mathcal{F}_2 = \left\{ \frac{0}{1}, \frac{1}{2}, \frac{1}{1} \right\},$$

$$\mathcal{F}_3 = \left\{ \frac{0}{1}, \frac{1}{3}, \frac{1}{2}, \frac{2}{3}, \frac{1}{1} \right\}.$$

2.1. Write out $\mathcal{F}_4, \mathcal{F}_5, \mathcal{F}_6$. Before reading beyond this first question, make as many observations about the Farey sequences as you can—you need not prove any of them. How can you find the elements of $\mathcal{F}_6$ just knowing $\mathcal{F}_5$? Are there any patterns? List all the observations you make and patterns you discover.

Before considering the arithmetical structure of $\mathcal{F}_N$, we take a brief detour to consider the size of $\mathcal{F}_N$. Let $\#(S)$ denote the number of elements (also known as the *order* or *cardinality*) of the set S. So for example, $\#(\mathcal{F}_3) = 5$.

2.2. Is $\#(\mathcal{F}_N)$ always a prime number? If so, explain why; if not, produce the smallest N for which $\#(\mathcal{F}_N)$ is not prime.

For a positive integer n, we define *Euler's phi function* (sometimes called *Euler's totient function*), $\varphi(n)$, to be the number of positive integers less than or equal to n and relatively prime to n. So, for example, $\varphi(9) = 6$ since there are exactly six numbers in the range from 1 to 9: $1, 2, 4, 5, 7, 8$ that are all relatively prime to 9.

LEMMA 2.3. *For a positive integer N, the number of elements in $\mathcal{F}_N$ having denominator N is equal to $\varphi(N)$.*

As an aside, here is a list of the values of the Euler phi function for the first 10 positive integers:

$$\varphi(1) = 1, \quad \varphi(2) = 1, \quad \varphi(3) = 2, \quad \varphi(4) = 2, \quad \varphi(5) = 4,$$

$$\varphi(6) = 2, \quad \varphi(7) = 6, \quad \varphi(8) = 4, \quad \varphi(9) = 6, \quad \varphi(10) = 4.$$

It is amusing to notice that each value of the Euler phi function above is attained at least twice (note, for example, that $\varphi(7) = \varphi(9) = 6$). In general, does each value attained by the Euler phi function occur at least twice? No one knows the answer, and thus this question remains open.

2.4. Find a formula for $\#(\mathcal{F}_N)$ in terms of the Euler phi function.

While there does not appear to be a simple, closed formula for the answer you've just found, the simple function $(3/\pi^2)N^2$ approximates the formula surprisingly well. For example, $\#(\mathcal{F}_{10}) = 33$ while $300/\pi^2 \approx 30.4$; $\#(\mathcal{F}_{100}) = 3045$ and $30000/\pi^2 \approx 3039.6$.

We now focus our attention on the main goal of this module: uncovering the structure of $\mathcal{F}_N$. Perhaps a few of the results ahead are some of the observations you made at the beginning of this module. First we consider two technical lemmas.

LEMMA 2.5. *Let $\frac{p}{q} < \frac{r}{s}$ be two rational numbers satisfying $ps - rq = -1$. Then for all positive integers λ and μ, it follows that*

$$\frac{p}{q} \leq \frac{\lambda p + \mu r}{\lambda q + \mu s} \leq \frac{r}{s}.$$

Remark. In what follows, it may be useful to note that the hypothesis of Lemma 2.5 could be expressed as

$$\det \begin{pmatrix} p & r \\ q & s \end{pmatrix} = -1.$$

LEMMA 2.6. *Let $\frac{p}{q} < \frac{r}{s}$ be two rational numbers satisfying $ps - rq = -1$. Suppose that $\frac{a}{b}$ is a rational number satisfying $\frac{p}{q} \leq \frac{a}{b} \leq \frac{r}{s}$. Then there exist nonnegative integers λ and μ such that*

$$a = \lambda p + \mu r \quad and \quad b = \lambda q + \mu s.$$

THEOREM 2.7. *Let $\frac{p}{q} < \frac{r}{s}$ be two rational numbers satisfying $ps - rq = -1$. Then a rational number $\frac{a}{b}$ is an element of the interval $\left[\frac{p}{q}, \frac{r}{s}\right]$ if and only if $a = \lambda p + \mu r$ and $b = \lambda q + \mu s$ for some nonnegative integers λ and μ.*

THEOREM 2.8. *Let $\frac{p}{q} < \frac{r}{s}$ be any two consecutive fractions in $\mathcal{F}_N$. Then*

$$\det \begin{pmatrix} p & r \\ q & s \end{pmatrix} = -1.$$

[⋆ ⋆ ⋆ HINT]

Module 3

Discoveries of Dirichlet and Hurwitz

Suppose we have a drawer filled with black socks and white socks. Of course if we, without looking, pull out three socks, we are guaranteed to have a pair of socks of the same color. This trivial observation is an example of the *Pigeonhole Principle*. Using this elementary principle we are able to prove a theorem of Dirichlet from 1842 which is one of the pillars of diophantine approximation. The theory of Farey fractions provides us with the machinery to give a different, more "effective" proof. We will explore both arguments here.

Dirichlet's result leads to many interesting and important questions and is the inspiration for much of diophantine analysis. In 1891, Hurwitz produced the *best possible* version of Dirichlet's result. In this module we'll also consider this important improvement.

3.1. THEOREM (The Pigeonhole Principle). *Suppose we have N boxes and $N + 1$ objects. If we were to place the objects in the boxes, then no matter how we placed the objects, there must exist a box containing more than one object.*

(Prove this result in *two ways*: first by contradiction and then by induction on N.)

LEMMA 3.9. *If x and y are positive integers, then at most one of the following inequalities can hold:*

$$\frac{1}{xy} \geq \frac{1}{\sqrt{5}}\left(\frac{1}{x^2} + \frac{1}{y^2}\right), \quad \frac{1}{x(x+y)} \geq \frac{1}{\sqrt{5}}\left(\frac{1}{x^2} + \frac{1}{(x+y)^2}\right).$$

[⋆⋆ HINT]

3.10 HURWITZ'S THEOREM. *Let α be an irrational number. Then there exist infinitely many rational numbers p/q such that*

$$\left|\alpha - \frac{p}{q}\right| < \frac{1}{\sqrt{5}q^2}.$$

[⋆ ⋆ ⋆ HINT]

Next we'll discover that Hurwitz's Theorem is best possible in that the result would be false if the number $\sqrt{5}$ were to be replaced by any larger value. To prove this assertion we need only to exhibit an α for which Hurwitz's inequality cannot be improved. An example of such a number is the *golden ratio* $\frac{1+\sqrt{5}}{2}$. We now let $\varphi = \frac{1+\sqrt{5}}{2}$ and write $\overline{\varphi} = \frac{1-\sqrt{5}}{2}$ for its conjugate.

3.11. (a) Show that for any rational a/b, it follows that

$$\left|\frac{a}{b} - \varphi\right|\left|\frac{a}{b} - \overline{\varphi}\right| = \frac{1}{b^2}\left|a^2 - ab - b^2\right|.$$

(b) Show that $a^2 - ab - b^2$ cannot equal zero (recall that $b \neq 0$).
(c) Deduce that

$$\left|\frac{a}{b} - \varphi\right|\left|\frac{a}{b} - \overline{\varphi}\right| \geq \frac{1}{b^2}.$$

(d) If for some positive real number m, $\left|\frac{a}{b} - \varphi\right| < \frac{1}{mb^2}$, then $\left|\frac{a}{b} - \overline{\varphi}\right| < \frac{1}{mb^2} + \sqrt{5}$.   [⋆ HINT]

THEOREM 3.12. *The constant $\sqrt{5}$ in Hurwitz's Theorem is best possible. That is, Theorem 3.10 does not hold if $\sqrt{5}$ is replaced by a larger value.*

$[\star \ \star \ \star \ \text{HINT}]$

The proof of Theorem 3.12 reveals that the constant $\sqrt{5}$ in Hurwitz's Theorem cannot be replaced with a larger value since the inequality with that smaller upper bound would no longer hold for $\alpha = \frac{1+\sqrt{5}}{2}$. However, suppose now that we restrict α to be an irrational number *not equal to* $\frac{1+\sqrt{5}}{2}$. Under this restriction, is it possible we can improve the constant occurring in Hurwitz's result? This question leads to an interesting collection of surprising and deep results. We will return to these questions in Module 6. Next, however, we will build up some powerful machinery for constructing the rational approximates p/q from Dirichlet's Theorem.

Big Picture Question. Do you think that the exponent of 2 on q is best possible? Do you believe there are examples of α's for which q^2 is best possible? Do you believe there are examples where q^2 is not best possible? What properties would these different classes of numbers possess? Just state your guesses without justification.

Further Challenge. Let r, s, and Q be integers satisfying $\max\{|r|, |s|\} > Q \geq 1$. Then there exist relatively prime integers p and q such that $1 \leq \max\{|p|, |q|\} \leq Q$ and

$$|rq - sp| \leq \frac{\max\{|r|, |s|\}}{Q + 1}.$$

then

$$\frac{p_n}{q_n} = x_0 + \cfrac{1}{x_1 + \cfrac{1}{x_2 + \cfrac{1}{\ddots + \cfrac{1}{x_n}}}}. \tag{4.2}$$

Caution. The "fraction" in (4.2) is purely a formal object—it may not be a rational number. In particular, notice that p_N and q_N need not be integers. In fact, they may be irrational.

For case of notation, we write $[x_0, x_1, \ldots, x_N]$ to denote the formal expression

$$x_0 + \cfrac{1}{x_1 + \cfrac{1}{x_2 + \cfrac{1}{\ddots + \cfrac{1}{x_N}}}}.$$

The next corollary gives an easy procedure for generating the p_n's and q_n's just knowing their predecessors.

COROLLARY 4.2. *Given the notation in Lemma 4.1, it follows that*

$$p_{N+1} = x_{N+1}p_N + p_{N-1} \quad and \quad q_{N+1} = x_{N+1}q_N + q_{N-1},$$

for all $N \geq 2$.

We now see that the p_n's and q_n's are related in a particularly simple way.

COROLLARY 4.3. *For all $N \geq 1$, $p_N q_{N-1} - p_{N-1} q_N = (-1)^{N+1}$.*

The quantities p_N/q_N can be expressed just in terms of x_0 and the q_n's.

LEMMA 4.4. *Given the notation in Lemma 4.1, for any integer $N \geq 0$,*

$$\frac{p_N}{q_N} = x_0 + \sum_{n=1}^{N} \frac{(-1)^{n-1}}{q_{n-1}q_n}.$$

If we further assume that the real numbers x_n are positive, then we see an interesting growth pattern in the sequence $\{p_n/q_n\}$.

COROLLARY 4.5. *If $x_i > 0$ for all $i > 0$, then*

$$\frac{p_0}{q_0} < \frac{p_2}{q_2} < \cdots < \frac{p_{2n}}{q_{2n}} < \cdots \quad \cdots < \frac{p_{2m+1}}{q_{2m+1}} < \cdots < \frac{p_5}{q_5} < \frac{p_3}{q_3} < \frac{p_1}{q_1}.$$

Remark. The previous inequality implies three different conclusions: the even-indexed terms are strictly increasing, the odd-indexed terms are strictly decreasing, and every even-indexed term is less than any odd-indexed term.

If we further restrict the x_n's to be *integers*, then we begin to capture some diophantine structure.

THEOREM 4.6. *If $a_0, a_1, \ldots$ are integers with $a_n > 0$ for all $n > 0$, then*

$$\lim_{N \to \infty} [a_0, a_1, \ldots, a_N] = \alpha,$$

for some irrational α. Moreover, for all $n \geq 0$,

$$\left| \alpha - \frac{p_n}{q_n} \right| < \frac{1}{q_n q_{n+1}}.$$

Suggestion. First show that the limit exists and the inequality holds. Then using the inequality, show that α must be irrational.

The expression $[a_0, a_1, \ldots]$ from Theorem 4.6 is called the (*simple*) *continued fraction expansion* of α. The integer a_n is called the *nth partial quotient* of α, and the rational number p_n/q_n is the *nth convergent* of α.

We now describe an algorithm for determining the continued fraction expansion for any given real α. In this direction we require a few basic notions. Given a real number α, we write $[\alpha]$ for the *integer part of α*, that is, the integer such that $[\alpha] \leq \alpha < [\alpha] + 1$.

4.7. Is the integer part of a real number unique?

We define the *fractional part of α*, denoted by $\{\alpha\}$, to be $\{\alpha\} = \alpha - [\alpha]$.

4.8. Find the integer and fractional parts of 3.14. Do the same for -4.13. *Handle the second number with care!*

Module 5

Enforcing the law of best approximates

For $\alpha \in \mathbb{R}$ and $Q \geq 1$ an integer, Dirichlet's Theorem states that there exists a rational number p/q such that $1 \leq q \leq Q$ and

$$\left| \alpha - \frac{p}{q} \right| \leq \frac{1}{q(Q+1)}. \tag{5.1}$$

Multiplying inequality (5.1) by q yields

$$|\alpha q - p| \leq \frac{1}{Q+1}.$$

Therefore we see that the integer p is easily determined by q: p is the nearest integer to αq. Thus we can let p fade into the background by defining the *distance to the nearest integer function*, $\| \ \|$, by

$$\|x\| = \min\{|x - n| : n \in \mathbb{Z}\}.$$

Dirichlet's Theorem can now be stated: There exists an integer q satisfying $1 \leq q \leq Q$ such that $\|\alpha q\| \leq (Q+1)^{-1}$. (See how we made the p disappear?)

In this module we ask, for an irrational number α, if there is an algorithm to find the complete list of increasing positive integers $\{q_1, q_2, \dots\}$ such that

$$\|\alpha q_1\| > \|\alpha q_2\| > \cdots > \|\alpha q_n\| > \cdots . \tag{5.2}$$

a rational number satisfying $1 \leq q \leq q_N$ *and* $(p, q) \neq (p_{N-1}, q_{N-1})$, $(p, q) \neq (p_N, q_N)$. *Then*

(i) $|\alpha q_0 - p_0| > |\alpha q_1 - p_1| > \cdots > |\alpha q_n - p_n| > \cdots,$

(ii) $|\alpha q - p| > |\alpha q_{N-1} - p_{N-1}|$.

Note. In the hypotheses of the theorem, by $(p, q) \neq (p_{N-1}, q_{N-1})$ we mean that the ordered pair (p, q) is not equal to the ordered pair (p_{N-1}, q_{N-1}).

[⋆ ⋆ ⋆ HINT]

From Lagrange's Theorem we have that the convergents form the complete list of best approximates. Moreover, it follows immediately from Theorem 4.6 or 4.13 that the convergents also satisfy $\left|\alpha - \frac{p}{q}\right| < \frac{1}{q^2}$. Is the converse true? That is, if the previous inequality is satisfied for *some* rational p/q, does that imply that the rational is a convergent of α? Although the answer turns out to be "no", if we tighten the inequality, the answer becomes "yes". This result was first discovered by Legendre.

LEMMA 5.10. *Let a and b be two distinct real numbers. Then*

$$ab < \frac{1}{2}(a^2 + b^2).$$

THEOREM 5.11. *Let p_{N-1}/q_{N-1} and p_N/q_N be consecutive convergents to α. Then at least one of those rationals will satisfy*

$$\left|\alpha - \frac{p}{q}\right| < \frac{1}{2q^2}.$$

[⋆ HINT]

5.12. LEGENDRE'S THEOREM. *Suppose that p and q are relatively prime integers with $q > 0$ and such that*

$$\left|\alpha - \frac{p}{q}\right| < \frac{1}{2q^2}.$$

Then $\frac{p}{q}$ is a convergent of α.

[⋆ HINT]

Big Picture Question. Why are the convergents truly the "best rational approximations" to α?

Further Challenge. Let $\alpha = [a_0, a_1, \ldots]$ be an irrational number, p_n/q_n denote its nth convergent, and α_n be its nth complete quotient. Then show that for each integer $n \geq 1$, and integer a, $1 \leq a \leq a_{n+1} - 1$,

$$\left| \alpha - \frac{ap_n + p_{n-1}}{aq_n + aq_{n-1}} \right| = \frac{\alpha_{n+1} - a}{(aq_n + q_{n-1})(\alpha_{n+1}q_n + q_{n-1})}.$$

These rational approximations are known as the *secondary convergents* of α.

a stunning result showing a deep connection between these constants and integer solutions to the diophantine equation

$$x^2 + y^2 + z^2 = 3xyz.$$

Once again we are sensing a link between diophantine inequalities and equations. In this module we will explore these themes and outline the basic results.

We wish to find small constants c such that

$$\left| \alpha - \frac{p}{q} \right| \leq \frac{c}{q^2}$$

has infinitely many rational solutions p/q. It is easy to see that the previous inequality is equivalent to

$$q\|\alpha q\| \leq c.$$

It thus follows that the optimal constant for a particular α is given by

$$\mu(\alpha) = \liminf_{q \to \infty} q\|\alpha q\|.$$

The value $\mu(\alpha)$ is often referred to as the *Markoff constant for* α. We begin with an exploration into the quantity $\mu(\alpha)$.

LEMMA 6.1. *Let* $\alpha = [a_0, a_1, a_2, \ldots]$ *be an irrational number and* $p_N/q_N = [a_0, a_1, \ldots, a_N]$ *denote its Nth convergent. Then for all* $N > 0$,

$$\frac{q_{N-1}}{q_N} = [0, a_N, a_{N-1}, \ldots, a_1].$$

LEMMA 6.2. *Let* $\alpha = [a_0, a_1, a_2, \ldots]$ *be an irrational number and* p_N/q_N *be its Nth convergent. Then for all $N > 0$,*

$$q_N\|\alpha q_N\| = \Big([a_{N+1}, a_{N+2}, a_{N+3}, \ldots] + [0, a_N, a_{N-1}, \ldots, a_1] \Big)^{-1}.$$

[⋆ HINT]

LEMMA 6.3. *For an irrational number* $\alpha = [a_0, a_1, a_2, \dots]$,

$$\mu(\alpha) = \liminf_{N \to \infty} \left([a_{N+1}, a_{N+2}, a_{N+3}, \dots] + [0, a_N, a_{N-1}, \dots, a_1] \right)^{-1}.$$

[⋆⋆ HINT]

Two irrational numbers are said to be *equivalent* if from some point onward, their continued fraction expansions agree. That is, α and β are equivalent if

$$\alpha = [a_0, a_1, \dots, a_{N-1}, a_N, a_{N+1}, a_{N+2} \dots]$$

and

$$\beta = [b_0, b_1, \dots, b_M, a_N, a_{N+1}, a_{N+2} \dots].$$

THEOREM 6.4. *Let* α *be an irrational, real number. Then*

$$\mu(\alpha) \leq \frac{1}{\sqrt{5}}.$$

Moreover, $\mu(\alpha) = 1/\sqrt{5}$ *if and only if* α *is equivalent to* $\frac{1+\sqrt{5}}{2}$.

Note. From Question 4.12, recall that $\frac{1+\sqrt{5}}{2} = [1, 1, 1, 1, \dots]$.

[⋆ ⋆ ⋆ HINT]

Theorem 6.4 gives us the complete list of irrational numbers for which the Hurwitz bound $1/\sqrt{5}$ is best possible. If we remove those values from consideration, then we can produce an improved upper bound.

THEOREM 6.5. *Let* α *be an irrational, real number not equivalent to* $\frac{1+\sqrt{5}}{2}$. *Then*

$$\mu(\alpha) \leq \frac{1}{\sqrt{8}}.$$

Moreover, $\mu(\alpha) = 1/\sqrt{8}$ *if and only if* α *is equivalent to* $1 + \sqrt{2}$.

Note. From Question 4.12, recall that $1 + \sqrt{2} = [2, 2, 2, 2, \dots]$.

[⋆ HINT]

We could continue the above process to find successively smaller Markoff constants as we whittle down the allowable values of α. The collection of all Markoff constants is called the *Markoff spectrum*.

In view of Theorem 6.10, we can view the solutions to (6.1) as a tree where two solutions are joined with an edge if they are neighbors. Here is a tree showing the first few Markoff numbers.

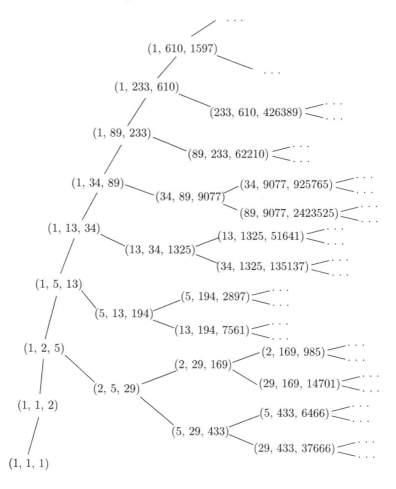

A tree showing the first few Markoff numbers and their neighbors

As an aside, we note that Hurwitz, in 1907, investigated general diophantine equations of the form

$$x_1^2 + x_2^2 + \cdots + x_N^2 = kx_1x_2\cdots x_N.$$

Among other things, he showed that $x^2 + y^2 + z^2 = kxyz$ has positive integer solutions if and only if k equals 1 or 3.

We close by describing the beautiful connection between the Markoff numbers and points in the Markoff spectrum above $\frac{1}{3}$. We'll just appreciate the result and not concern ourselves with its proof (see [7] for a proof). Recall that $\mu(\alpha)$ denotes the Markoff constant for α, that is, $\mu(\alpha) = \liminf_{q \to \infty} q\|\alpha q\|$.

MARKOFF'S THEOREM. *The Markoff spectrum above $1/3$ consists precisely of numbers of the form $m/\sqrt{9m^2 - 4}$, where m is a positive integer such that there exist positive integers k and l, with $\max\{k,l\} \leq m$ satisfying*

$$k^2 + l^2 + m^2 = 3klm.$$

COROLLARY 6.11. *The largest accumulation point of the Markoff spectrum is $\frac{1}{3}$.*

6.12. Using the solutions given in the Markoff tree above, verify the first few entries in the table of Markoff constants.

We now wonder what can be said about the Markoff spectrum *below* $1/3$. The answer is that the spectrum gets extremely complicated: there are uncountably many points with various gaps. The smallest gap ends at *Freiman's number*

$$\frac{153640040533216 - 19623586058\sqrt{462}}{693746111282512} = 0.220856369\cdots.$$

Every real positive number less than Freiman's number is in the Markoff spectrum!

Big Picture Question. Let $\varepsilon > 0$. Show that there exist irrational numbers α satisfying $\mu(\alpha) < \varepsilon$. Do you think there are α's for which $\mu(\alpha) = 0$?

We say an irrational number α is *badly approximable* if there exists a constant $c = c(\alpha) > 0$ such that for *all* rational numbers p/q,

$$\frac{c}{q^2} < \left| \alpha - \frac{p}{q} \right|. \tag{7.2}$$

We emphasize that c is a constant depending only on α and not on the individual p/q under consideration. The same c must work for *all* rational numbers.

7.1. Show that the exponent of 2 in the corollary to Dirichlet's Theorem (Corollary 3.5) cannot be improved if α is badly approximable. Using your results regarding Hurwitz's Theorem or Markoff's work, give an example of a badly approximable number and prove that your example is badly approximable. Your example will conclusively demonstrate that badly approximable numbers exist!

LEMMA 7.2. *If* $\alpha = [a_0, a_1, \dots]$ *is an irrational number, then for all* $N \geq 1$,

$$\frac{1}{(a_{N+1} + 2)q_N^2} \leq \left| \alpha - \frac{p_N}{q_N} \right| \leq \frac{1}{a_{N+1}q_N^2}.$$

We now come to a simple, necessary and sufficient condition on the partial quotients to ensure a number is badly approximable.

THEOREM 7.3. *An irrational number is badly approximable if and only if its partial quotients are bounded.*

Warning. In showing a number is badly approximable, one must show that an inequality of the form (7.2) holds for *all* rational numbers, not just for the convergents p_n/q_n.

THEOREM 7.4. *Let* α *be an irrational, real number and let* $\mu(\alpha)$ *denote the Markoff constant of* α *as defined in Module 6. Then* α *is a badly approximable number if and only if* $\mu(\alpha) > 0$.

We note that Theorem 7.4 answers part of the Big Picture Question from the previous module.

7.5. Compute the continued fraction expansions of the following numbers: $\sqrt{2}$, $\frac{4+\sqrt{3}}{7}$, $1 + \sqrt{5}$, and $\sqrt{29}$. Using your results, make a conjecture about the continued fraction expansion for certain irrational numbers. Find the continued fraction expansion for $\sqrt{N^2 + 1}$, where $N \geq 1$ is an integer.

An irrational number α is said to be a *quadratic irrational* if it is a root of a quadratic equation of the form $ax^2 + bx + c = 0$, where a, b, and c are all integers.

THEOREM 7.6. *If α has a continued fraction expansion that is eventually periodic, then α is a quadratic irrational.*

[$\star\star$ HINT]

Does the converse of Theorem 7.6 hold? In an attempt to answer this question, we consider a technical lemma.

LEMMA 7.7. *If α is a quadratic irrational, then it can be written in the form $\alpha = (r + \sqrt{d})/s$, where d is a positive integer that is not a perfect square, and r and s are integers such that $s \mid d - r^2$. Furthermore, if there are integers d, r, and s such that $\alpha = (r + \sqrt{d})/s$ but $s \nmid d - r^2$, then there exist integers D, R, and S such that $\alpha = (R + \sqrt{D})/S$ and $S \mid D - R^2$.*

Using Lemma 7.7, we are able to give another method to compute the continued fraction expansion of a quadratic irrational. This algorithm is due to L. Euler. We recall that $[x]$ denotes the integer part of x.

THEOREM 7.8. *Let α be a quadratic irrational. Let $\alpha_0 = \alpha$, $a_0 = [\alpha_0]$. Using Lemma 7.7, we can write $\alpha_0 = (r_0 + \sqrt{d})/s_0$, where d is not a perfect square and $s_0 \mid d - r_0^2$. Let $\alpha_{n+1} = 1/(\alpha_n - a_n)$ and $a_{n+1} = [\alpha_{n+1}]$. Finally, for $n \geq 0$, we define*

$$r_{n+1} = a_n s_n - r_n \quad and \quad s_{n+1} = \frac{d - r_{n+1}^2}{s_n}.$$

Then for all $n \geq 0$, it follows that:

(*i*) $\alpha = [a_0, a_1, a_2, \dots]$;
(*ii*) r_n and s_n are integers with $s_n \neq 0$;
(*iii*) $s_n \mid d - r_n^2$;
(*iv*) $\alpha_n = (r_n + \sqrt{d})/s_n$, and thus $a_n = [(r_n + \sqrt{d})/s_n]$.

Big Picture Question. Why is the continued fraction expansion of a real number so sympathetic to quadratic irrationals?

(See Appendix 1 for some commentary on this question.)

Further Challenge. Give an upper bound for the length of the period of the continued fraction expansion for the quadratic irrational $\alpha = (r + \sqrt{d})/s$.

Module 8

Solving the alleged "Pell" equation

Many young aspiring mathematicians dream about having a mathematical object named after them. Do you have to be a mathematical superstar to see your name in (mathematical) lights? The answer is "yes", *or* just have a mathematical superstar accidently credit you with someone else's work—your name may very well stick! This module is the story of a diophantine equation incorrectly attributed to Pell. Let d be a fixed positive integer that is not a perfect square. Then the *Pell equation* is defined by

$$x^2 - dy^2 = \pm 1.$$

John Pell was a notable 17th-century English mathematician who had essentially no connection with the previous equation. However the great mathematician Leonhard Euler mistakenly credited Pell with some results involving this equation that, in actuality, were due to Lord Brouncker. Although most people are aware of Euler's error, Pell's name remains associated with the equation to this very day (on this very page).

In this module we will find all the integer solutions to the Pell equation and discover some surprising connections between the solutions and the continued fraction expansion for $\sqrt{d}$. As $\sqrt{d}$ is a quadratic irrational, we know that its continued fraction expansion will

where the string $(a_1, a_2, a_3, \ldots, a_{N-1})$ *is a palindrome.*

As an illustration, we note that

$$\sqrt{61} = \left[7, \overline{1, 4, 3, 1, 2, 2, 1, 3, 4, 1, 14}\right].$$

We now turn our attention to the Pell equation. In this direction, we first recall Euler's algorithm for computing the continued fraction expansion of quadratic irrationals (Theorem 7.8):

EULER'S THEOREM.   *Let* α *be a quadratic irrational. If* $\alpha_0 = \alpha$, $a_0 = [\alpha_0]$, *then* $\alpha_0 = (r_0 + \sqrt{d})/s_0$, *where* d *is not a perfect square and* $s_0 | d - r_0^2$. *Let* $\alpha_{n+1} = 1/(\alpha_n - a_n)$ *and* $a_{n+1} = [\alpha_{n+1}]$. *Finally, for* $n \geq 0$, *we define*

$$r_{n+1} = a_n s_n - r_n \quad and \quad s_{n+1} = \frac{d - r_{n+1}^2}{s_n}.$$

Then for all $n \geq 0$, *it follows that:*
   (i) $\alpha = [a_0, a_1, a_2, \ldots]$;
   (ii) r_n *and* s_n *are integers with* $s_n \neq 0$;
   (iii) $s_n | d - r_n^2$;
   (iv) $\alpha_n = (r_n + \sqrt{d})/s_n$, *and thus* $a_n = [(r_n + \sqrt{d})/s_n]$.

Using the results of this module and Euler's algorithm, we are led to the following technical lemma.

LEMMA 8.9. *Let* d *be a positive integer, not a perfect square, and let* $\alpha = \sqrt{d} + \left[\sqrt{d}\right]$. *Suppose that* N *is the shortest period in the continued fraction of* α. *Then given the notation in Euler's Theorem, the following all hold:*
   (i) $\alpha_1, \ldots, \alpha_{N-1}$ *are all distinct from* α_0.
   (ii) $\alpha_m = \alpha_0$ *if and only if* $m = tN$ *for some integer* $t \geq 0$.
   (iii) *For all integers* $t \geq 0$, $r_{tN} - s_{tN}\left[\sqrt{d}\right] = (s_{tN} - 1)\sqrt{d}$.
   (iv) $s_m = 1$ *if and only if* $m = tN$ *for some integer* $t \geq 0$.
   (v) *There do not exist indices* m *such that* $s_m = -1$.

$[\star\star \text{ HINT}]$

LEMMA 8.10. *Let d be a positive integer, not a perfect square, and p_n/q_n denote the nth convergent of $\sqrt{d}$. Given the notation in Euler's Theorem, for all $n \geq 0$,*

$$\sqrt{d} = \frac{(r_{n+1} + \sqrt{d})p_n + s_{n+1}p_{n-1}}{(r_{n+1} + \sqrt{d})q_n + s_{n+1}q_{n-1}}. \tag{8.1}$$

[⋆ HINT]

We are finally ready to consider solutions to certain diophantine equations.

THEOREM 8.11. *Let d be a positive integer, not a perfect square. Given the notation in Lemma 8.10, for all $n \geq 0$,*

$$p_n^2 - dq_n^2 = (-1)^{n-1}s_{n+1}.$$

[⋆ HINT]

We are now in a position to say something about the solutions to the Pell equation.

COROLLARY 8.12. *Let d be a positive integer, not a perfect square, and suppose that N is the shortest period in the continued fraction of $\sqrt{d}$. Then for all $t \geq 0$,*

$$p_{tN-1}^2 - dq_{tN-1}^2 = (-1)^{tN}.$$

Moreover, if

$$p_m^2 - dq_m^2 = \pm 1,$$

then m must equal $tN - 1$ for some integer $t \geq 0$.

Remark. We define $p_{-1} = 1$ and $q_{-1} = 0$.

Using Corollary 8.12 and Legendre's Theorem (Theorem 5.12), we can produce the definitive word on the solutions to the Pell equation:

THEOREM 8.13. *Let d be a positive integer, not a perfect square, and suppose that N is the shortest period in the continued fraction of $\sqrt{d}$. Then the positive integer solutions (x, y) to the Pell equation*

$$x^2 - dy^2 = \pm 1$$

Exactly 30 years after Liouville's result, Cantor developed his deep and revolutionary theory of sets. We'll close this module with some diophantine consequences of Cantor's work.

A number α is said to be *algebraic* if it is a root of a polynomial of the form

$$f(z) = a_N z^N + a_{N-1} z^{N-1} + \cdots + a_1 z + a_0,$$

where $a_n \in \mathbb{Z}$ and $a_N \neq 0$. The *minimal polynomial* associated with an algebraic number α is the polynomial with (relatively prime) integer coefficients, $a_N > 0$, having the smallest degree for which α is a root. In other words, it is the *irreducible polynomial* associated with α—the polynomial that cannot be factored into two polynomials with integer coefficients each having degrees less than N. For example, $\sqrt{3}$ is algebraic because it is a root of the polynomial $h(z) = z^4 - 9$. However, $h(z)$ is not the minimal polynomial for $\sqrt{3}$ since $h(z)$ can be factored further: $h(z) = \left(z^2 - 3\right)\left(z^2 + 3\right)$. The minimal polynomial for $\sqrt{3}$ is in fact $f(z) = z^2 - 3$. Notice that $f(z)$ is irreducible. We define the *degree* of an algebraic number to be the degree of its minimal polynomial. So the degree of $\sqrt{3}$ is 2.

A number that is not algebraic is called *transcendental*. Certainly, as we've seen in the previous paragraph, algebraic numbers exist. Do transcendental numbers exist? This question was one of the major open problems from 19th-century mathematics. This question was finally answered by Liouville who used techniques from diophantine approximation.

Liouville's Theorem provides a measure for how well algebraic numbers can be approximated by rational numbers that are not too complicated. We edge our way up to Liouville's result slowly.

9.1. Suppose that $f(x)$ is a differentiable function and a and b are real numbers satisfying $|a - b| \leq 1$. Use the Mean Value Theorem to give a bound of the form

$$|f(b) - f(a)| \leq C(f, a)|b - a|,$$

where $C(f, a)$ is a constant that just depends upon the function f and the number a, but *not* on b.

9.2. Suppose that $f(x)$ is a nonconstant polynomial. If $0 < |a - b| \leq 1$, then the constant $C(f, a)$ found above is nonzero.

LEMMA 9.3. *Let $f(x)$ be a polynomial of degree d, with integer coefficients. Suppose that the rational number p/q is not a root of f. Then*

$$\frac{1}{q^d} \leq \left| f\left(\frac{p}{q}\right) \right|.$$

LEMMA 9.4. *Suppose that α is a real number and p/q is a rational number such that $|\alpha - p/q| > 1$. Then for any positive integer d,*

$$\frac{1}{q^d} \leq \left| \alpha - \frac{p}{q} \right|.$$

9.5. LIOUVILLE'S THEOREM. *Let α be a real algebraic number of degree $d \geq 2$. Then there exists a constant $c = c(\alpha) > 0$ such that for all rational numbers p/q,*

$$\frac{c}{q^d} < \left| \alpha - \frac{p}{q} \right|.$$

$[\star\ \star\ \star\ \text{HINT}]$

9.6. Show that Liouville's Theorem implies that all rational numbers near algebraic numbers are complicated. Specifically, show that if α is an algebraic number and p/q is a rational number that is very, very close to α, then the height of p/q (*i.e.*, q) must be very large.

9.7. Suppose that α is a transcendental number (note that at this point we do not know if such animals exist). Does Liouville's Theorem imply anything about how well α can be approximated by rationals having small height?

9.8. What is the contrapositive of Liouville's Theorem? What does that statement imply about numbers that have amazing approximations by rationals of small height?

An important consequence of Liouville's celebrated result is the existence of transcendental numbers.

It is an open question as to whether or not the apparent lack of pattern continues. In particular, the question of whether π is a badly approximable number or not remains open. The general consensus is that π is not badly approximable, but for now that is nothing more than a guess. Finally, you are welcome to ask the same types of questions for various other numbers, for example $e+\pi$. Unfortunately, no one has been able to prove that $e+\pi$ is even an *irrational* number. We'll leave this as a bonus problem.

Big Picture Question. What is the basic idea behind Liouville's argument? Numbers having similar diophantine structure as that of the α from Theorem 9.9 are known as *Liouville numbers*. All Liouville numbers are transcendental, and the proof of Theorem 9.9 can be adapted to show that these numbers are transcendental. What should be the definition of a *Liouville number*?

(See Appendix 1 for some commentary on this question.)

Further Challenge. Suppose that $\alpha = [a_0, a_1, a_2, \ldots]$ is an irrational number with the property that there is an infinite *subsequence* of partial quotients $\{a_{n_i}\}$ that grows incredibly fast. In particular, suppose that $a_{n_i} \geq q_{n_i-1}^{n_i}$, where q_n denotes the denominator of the nth convergent of α. Prove that α must be transcendental.

Module 10

Roth's stunning result and its consequences

Liouville's Theorem gives a lower bound for how well an algebraic number can be approximated by rationals. In particular, for an algebraic number α of degree $d \geq 2$, there exists a constant $c = c(\alpha) > 0$ such that for all rational numbers p/q,

$$\frac{c}{q^d} < \left| \alpha - \frac{p}{q} \right|.$$

In this module we ask if Liouville's inequality can be sharpened. That is, can we replace the lower bound by a larger quantity and yet still have the inequality hold for all p/q? In particular, can we replace the exponent d by a *smaller* quantity? Here we will explore these questions and see that after many improvements by many important mathematicians, in 1955 K. Roth produced the best possible result. His work was so important that he was awarded the Fields Medal (the mathematical version of the Nobel Prize) that same year. Although we will not prove his important result, we will outline a strategy for the proof and then consider consequences of his theorem within the context of diophantine equations.

10.1. Do there exist α's for which Liouville's inequality can be improved so that the lower bound would have the form c/q^λ, where $c = c(\alpha)$ is a positive constant and λ is some real number such that

The construction of the polynomial is easy—we just let $f(x)$ be the minimal polynomial for α and use the Mean Value Theorem to compare $f(p/q)$ with $|\alpha - p/q|$. We also know that $f(x)$ cannot vanish at the rational p/q. In fact we are able to show that $f(p/q)$ is bounded away from zero by at least $1/q^d$. Putting these pieces together, we can prove Liouville's Theorem.

Turning our attention to Roth's Theorem, we first observe that from 10.5, to prove Roth's Theorem it is enough to prove Theorem 10.4. Roth proves Theorem 10.4 by contradiction: he assumes there are infinitely many rational numbers p/q that satisfy (10.1). In particular, he knows there exist rationals

$$\frac{p_1}{q_1}, \frac{p_2}{q_2}, \dots, \frac{p_M}{q_M} \tag{10.2}$$

all satisfying (10.1) such that the denominators are growing very rapidly (*i.e.*, q_{m+1} is larger than q_m^2), and where M is any large integer satisfying $M \geq 2\varepsilon^{-2}\log(2d)$. The rationals in (10.2) can be thought of as *excellent* rational approximations to α. Next Roth constructs a polynomial with integer coefficients *in M variables*, $P(x_1, x_2, \dots, x_M)$, that satisfies some incredible properties. Here are the basic properties of P.

PROPERTY 1. The coefficients of $P(x_1, x_2, \dots, x_M)$ are fairly small, while the degrees in the individual variables, x_m, are increasing exponentially (that is, the degree in x_m is exponentially larger than the degree of x_{m-k}).

PROPERTY 2. The polynomial $P(x_1, x_2, \dots, x_M)$ vanishes strongly at $(\alpha, \alpha, \dots, \alpha)$. That is, not only does $P(\alpha, \alpha, \dots, \alpha) = 0$, but *many* partial derivatives vanish at this point as well. Thus, in some sense, the graph of the function would be extremely flat near the point $(\alpha, \alpha, \dots, \alpha)$.

Here are the two major meta-ideas of Roth's proof.

IDEA 1. Recall that we selected the rational M-tuple $(p_1/q_1, p_2/q_2, \dots, p_M/q_M)$ so that it is very close to $(\alpha, \alpha, \dots, \alpha)$ (each rational satisfies (10.1)). This observation, together with the fact that P is very flat around $(\alpha, \alpha, \ldots, \alpha)$, implies that P must vanish *fairly* strongly at the point $(p_1/q_1, p_2/q_2, \ldots, p_M/q_M)$.

IDEA 2. This is the key idea: Roth proves that if the degrees of P increase exponentially, then P cannot vanish *too* strongly at $(p_1/q_1, p_2/q_2, \ldots, p_M/q_M)$. This result is known as Roth's Lemma and contradicts the conclusions from the previous idea. Thus our assumption was false and hence there are only finitely many rational solutions to (10.1).

In some vague sense, we can see reflections of the strategy used by Roth in the elementary argument of Liouville's Theorem. To prove Roth's Theorem we

- Construct a polynomial P that vanishes strongly at $(\alpha, \alpha, \ldots, \alpha)$ and thus fairly strongly at $(p_1/q_1, p_2/q_2, \ldots, p_M/q_M)$.
- Show that P cannot vanish too strongly at $(p_1/q_1, p_2/q_2, \ldots, p_M/q_M)$.

These two steps lead to the contradiction. Before closing this discussion, we make one last observation. In Liouville's Theorem we were able to explicitly produce the constant $c(\alpha)$. In Roth's argument we are only able to conclude that the constant $c(\alpha, \varepsilon) > 0$ *exists*, since the proof gives no means by which to compute it. Thus Roth's Theorem is referred to as an *ineffective result* in that we know the constant exists, but we have no way of finding it. The basic idea behind the ineffectivity revolves around the fact that either α has excellent approximations or not. If it does not, then the constant can be explicitly given. However if α does have excellent approximations, then the constant depends on α, ε, *and* the excellent approximations. Unfortunately, in the course of the proof we never find out if an excellent approximation exists or not. Thus we know the constant exists, but we do not know its value. There is great interest among researchers in this area to produce *effective* versions of Roth's Theorem. We'll see the value of such effective results as we consider an application of this theory.

The Thue-Siegel-Dyson-Roth type theorems have important implications to the theory of diophantine equations. As an illustration, let's consider the following result due to A. Thue from 1909.

LEMMA 11.1. *There is a one-to-one correspondence between Pythagorean triples, (x, y, z), with $z \neq 0$, and rational points on the unit circle*

$$X^2 + Y^2 = 1. \tag{11.1}$$

Within the proof of Lemma 11.1, we see that finding *integer* solutions to $x^2 + y^2 = z^2$ with $z \neq 0$ is equivalent to finding *rational* solutions to $X^2 + Y^2 = 1$. Thus it is enough to find all rational points on the unit circle. We begin by noting that $(-1, 0)$ is one such solution to (11.1).

LEMMA 11.2. *If (x, y) is rational solution to (11.1), other than $(-1, 0)$, then the line through $(-1, 0)$ and (x, y) has rational slope. Conversely, for any rational number m, the line passing through $(-1, 0)$ with slope m intersects the unit circle in two rational points.*

Thus we are finding a one-to-one correspondence between the rational numbers (the slopes of lines) and rational points of the curve. This link between arithmetic (rational points) and geometry (slopes of lines) is at the heart of arithmetic geometry.

LEMMA 11.3. *Suppose that $m = s/r$. Then the line of slope m passing through $(-1, 0)$ also intersects the unit circle at*

$$\left(\frac{r^2 - s^2}{r^2 + s^2}, \frac{2rs}{r^2 + s^2} \right).$$

If we multiply through by the denominator, we see that

$$(r^2 - s^2)^2 + (2rs)^2 = (r^2 + s^2)^2.$$

We are now ready to describe all primitive Pythagorean triples.

THEOREM 11.4. *There are infinitely many primitive Pythagorean triples (x, y, z). Moreover they are all given by*

$$x = s^2 - r^2, \quad y = 2rs, \quad z = s^2 + r^2,$$

where r and s are relatively prime positive integers such that $s > r > 0$ and either r or s is even.

Why did the connection between lines and curves work so nicely? The key is that the degree of the polynomial in equation (11.1) is two, and thus if a line intersects the curve at one point, then it must intersect the curve at a second point (with the understanding that the line may be tangent to the curve). Thus an analysis of lines having rational slopes leads to rational points on the curve. This observation illustrates a connection between the degree of a polynomial equation and the structure of its rational solutions. We'll explore this connection a bit further.

Let $f(x, y)$ be a polynomial in two variables with integer coefficients. The solutions in $\mathbb{R}^2$ to $f(x, y) = 0$ form some curve in the plane. The curve is one dimensional over the real numbers in the sense that every point of the curve has a neighborhood which looks like a piece of the real line (except that it is curved). Actually the previous statement is not quite accurate. It's possible that the curve might pass through itself and thus form an $\times$ or a cusp. The point where the curve crosses itself or the point of a cusp is called a *singularity*. To avoid this annoying complication, let's assume that the curve is *nonsingular*, meaning that the curve does not cross itself and contains no cusp points. (Technically, a curve is nonsingular at a point if the partial derivatives of $f(x, y)$ do not both vanish there.)

The complex solutions $(x, y) \in \mathbb{C}^2$ to $f(x, y) = 0$ form a curve over the complex numbers. Such a curve over $\mathbb{C}$ is called a *Riemann surface*. This curve is one dimensional over $\mathbb{C}$ which means that around every point there is a neighborhood that looks like a piece of the complex plane (again, we are assuming the curve is nonsingular). Thus the "curve" actually looks like a surface. From now on, let's assume that all the surfaces we consider here are nonsingular.

11.5. Let S be a sphere with the north pole removed and suppose it sits on a plane (tangent to its south pole). By considering line segments emanating from the north pole, prove that there is a one-to-one correspondence between the points on the plane and the points of S. This one-to-one correspondence is known as *stereographic projection*.

Thus there is a way of "pulling up the edges at infinity" of the plane to make an infinite "bowl" of sorts. The open top of the bowl

terms of the degree of the polynomial. If the degree of the polynomial is d, then the genus $g = \binom{d-1}{2} = (d-1)(d-2)/2$.

11.10. What is the genus of the surface associated with $x^2 + y^2 = 1$?

In general, the basic question of finding rational solutions to genus zero surfaces has been answered. Once we have one rational solution, we can adopt the same method as in Theorem 11.4 to show that there are infinitely many rational solutions, and we can even write down an explicit formula that generates all of them.

11.11. Find all rational points on the ellipse $x^2 + 5y^2 = 1$.

There are many deep and important conections between the genus of a curve and the rational points on the curve. We have already remarked that if $g = 0$, then the basic problem can be solved completely. In 1983 Faltings proved an extremely difficult conjecture of Mordell made in 1923 stating that if the genus is two or more (that is, if the degree of the polynomial is 4 or more), then there can be at most a *finite* number of rational points on the curve. Three years later, Faltings was awarded the Fields Medal for his ground-breaking work in this direction.

Although there have been great breakthroughs in arithmetic geometry, many basic questions remain unanswered. For example, no one knows a procedure to find *all* rational points on curves of a given genus $g \geq 2$. Is there a bound on the size of the numerators and denominators (*i.e.*, the height) of the coordinates of the solutions? Is there a uniform bound for the number of rational points a curve of genus 2 can have? These difficult, fundamental questions remain open.

If the genus is exactly one (so the degree of the polynomial is equal to 3), then it is possible that there may be no rational points on the curve. If, however, there is a rational point on the curve, then the curve is called an *elliptic curve*. We describe these curves and their connections with abstract algebra in the next module.

Big Picture Question. The Riemann surface of $x^2 + y^2 = 1$, when viewed in the projective completion of $\mathbb{R}^2$, is a sphere. The graph of $x^2 + y^2 = 1$ in the plane is a circle which represents a "level curve" or slice of the sphere. Using methods from calculus, produce a careful sketch of the curve $y^2 = x^3 - x$. If you now view this curve in the projective completion of $\mathbb{R}^2$, what would it look like? The curve, which is an example of an elliptic curve, is a "level curve" of some surface. What do you think the surface is? (Don't forget about the points at infinity.)

(See Appendix 1 for some commentary on this question.)

Further Challenge. Let (x, y, z) be a primitive Pythagorean triple. Prove that one of the integers in the triple is divisible by 3, one is divisible by 4 and one is divisible by 5.

Remark. The numbers 3, 4, and 5 need not necessarily divide *distinct* elements of the triple. For example, note that 3, 4, and 5 all divide the same element of the primitive Pythagorean triple $(11, 60, 61)$.

An elliptic curve is *nonsingular* if the curve never crosses itself and contains no cusps. Alternatively, a curve is nonsingular if its partial derivatives do not both vanish at the same point on the curve.

12.2. Let n be a fixed, positive integer. Consider the curves given by

$$y^2 = x((x-1)^2 + n), \qquad y^2 = x(x-1)^2, \qquad y^2 = x((x-1)^2 - n).$$

Which of these are elliptic curves and why? Which of these are singular and which are nonsingular? Give a *rough* sketch of each curve in $\mathbb{R}^2$.

For the remainder of this module, let's assume that our curve is nonsingular. We will also assume that the equation associated with our elliptic curve is of the form

$$y^2 = x^3 + ax + b, \tag{12.1}$$

where a and b are rational numbers. It can be shown that every elliptic curve has an equation which is, in some sense, "equivalent" to equation (12.1). It certainly is reasonable to wonder, "Why are these curves called elliptic?" Even though elliptic curves are not ellipses, they do arise within the context of ellipses. If someone had the impish desire to find the arc length of an *ellipse*, that person would end up being face-to-face with an integral having an "elliptic curve" in its denominator (and actually under a square root)—hence its name.

Suppose we are given two points on an elliptic curve. How could we use the two to generate a third point on the curve? One way to proceed is through the geometry of the curve itself.

12.3. Let P and Q be two distinct points on an elliptic curve, and let L be the line containing P and Q. Prove that if L is not a vertical line, then it must intersect the curve at another point. Furthermore, show that if both P and Q are rational points, then the third point of intersection of the curve and the line L is also a rational point.

Let $\mathcal{C}$ be the elliptic curve associated with equation (12.1) and let $E_{\mathcal{C}}(\mathbb{Q})$ (or just $E(\mathbb{Q})$) be the set of all rational points on $\mathcal{C}$. Inspired by the answers to 12.3, let's define a binary operation on $E(\mathbb{Q})$ and show that under this binary operation, $E(\mathbb{Q})$ has some interesting algebraic structure.

Suppose that P and Q are elements of $E(\mathbb{Q})$ having *distinct* x coordinates. Then from 12.3 we see that there exists a unique element $R' \in E(\mathbb{Q})$ so that P, Q and R' lie on the same straight line. Let R be defined by $R = -R'$. So R is the reflection of R' about the x-axis.

12.4. Verify that $R \in E(\mathbb{Q})$.

We now define *addition on* $E(\mathbb{Q})$ by $P + Q = R$ (see the figure below where the elliptic curve depicted is $y^2 = x^3 - x$).

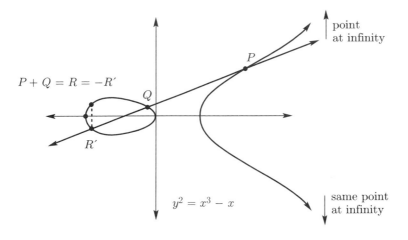

Now what if P and Q are distinct points of $E(\mathbb{Q})$ but have the *same x coordinate*? To find $P+Q$ we would look at the line connecting P and Q, and see where it intersects the curve. But that line would be vertical (its slope is ∞) and thus it does not cross the curve at any other point. This situation appears problematic. But we are saved if we add in one point at infinity! That is, if we consider the *projective completion*. In particular, we add in the point at infinity where all vertical lines intersect. It turns out that this point at infinity is *on the curve* (notice that in the previous figure, the curve is heading up and down toward our vertical point at infinity). In fact it can be shown that this point at infinity is a *rational point* on the curve. Thus this

the end of our brief journey through the ideas of diophantine geometry. We close our overview with some questions in order for you to have a chance to think about these themes on your own.

12.7. Let $\mathcal{C}$ be the curve associated with $y^2 = x^3 + 4x$. Show that $P = (2, 4) \in E(\mathbb{Q})$. Prove that the tangent line of $\mathcal{C}$ at P contains $(0, 0)$. Now determine the order of P in the group $E(\mathbb{Q})$. That is, how many times must P be added to itself in order to yield the identity element?

12.8. Let $\mathcal{C}$ be the curve associated with $y^2 = x^3 + ax + b$ and assume that $\mathcal{C}$ is nonsingular. Find all points in $E(\mathbb{Q})$ of order 2. [REMARK: There are two cases to consider.] Next explain geometrically how you could set out to find all points of order 4 in $E(\mathbb{Q})$.

12.9. At the beginning of this module we **boldly** stated that the only rational points on the elliptic curve $\mathcal{C}$ associated with $y^2 = x^3 - 4x$ are $(2, 0), (0, 0), (-2, 0)$ and the point at infinity. Verify this claim. That is, prove that $E(\mathbb{Q}) = \{\infty, (2, 0), (0, 0), (-2, 0)\}$.

[⋆ ⋆ ⋆ HINT]

12.10. What well-known group is isomorphic to the group $E(\mathbb{Q})$ described in 12.8? Justify your answer.

Note. This question is not "trivial"; there are a few possibilities and you need to explain why the group is the one you claim.

12.11. Prove that the diophantine equation $y^2 = x^4 + x^3 + x^2 + x + 1$ has integral solutions $(-1, \pm 1)$, $(0, \pm 1)$, $(3, \pm 11)$, and no others.

[⋆ ⋆ ⋆ HINT]

Big Picture Question. How do the theorems of Mordell and Mazur fit into the diophantine landscape of deducing facts about rational solutions to diophantine equations?

Further Challenge. Let r be a rational number, $r \neq 0$, $r \neq \frac{1}{4}$. Then show that the point (r, r) is a point of order four on the nonsingular elliptic curve

$$y^2 = x^3 - (2r - 1)x^2 + r^2 x.$$

Module 13

The geometry of numbers

In the two previous modules we've seen a hint of how geometry and algebra come together to deduce deep results about certain diophantine equations. In this module we will explore another powerful interface between geometry and algebra known as the geometry of numbers. Here the bridge between geometry and algebra provides interesting results in diophantine approximation. The subject of geometry of numbers was developed, essentially single handedly, by Minkowski in the late 19th century.

The basic study of geometry of numbers revolves around the desire to understand the interaction between a set possessing algebraic structure and another set exhibiting geometric structure. The most fundamental question one could ask is, Do these sets contain a point in common? The subject really starts there and then blossoms into a truly beautiful theory that has far-reaching applications and ramifications to many diverse areas, including number theory, coding, coverings, and packings.

We begin by investigating sets that possess algebraic structure. In particular, we study lattices. A terse definition of a *lattice* Λ is a subset of $\mathbb{R}^N$ that is a $\mathbb{Z}$-module generated by N $\mathbb{R}$-linearly independent vectors. A less terse, but perhaps more comprehensible

where $\chi_S(x)$ represents the *characteristic function for the set* S, that is,

$$\chi_S(x) = \begin{cases} 1 & \text{if } x \in S, \\ 0 & \text{if } x \notin S. \end{cases}$$

13.5. Let R be the rectangle having vertices $(0,0)$, $(0,b)$, $(a,0)$, (a,b). Graph $\chi_R(\vec{x})$ (note that in this case, $N = 2$). Let C be the disk centered at $(1,1)$ of radius 1. Graph $\chi_C(\vec{x})$. Using the definition of N-dimensional volume and your two graphs, find the area (*i.e.*, 2-dimensional volume) of R and C.

There are two technical points that we'll mention very quickly. First, not every set in $\mathbb{R}^N$ has a volume. A set that has volume is called a *measurable set*. Also, the integral above really denotes a "Lebesgue integral". These remarks and the measure theory that lies behind them allow one to integrate very complicated functions. Happily, for our purposes here, we need not worry about any of these subtleties. We may treat the previous integral as a good, old-fashioned Riemann integral and not worry about the measurablity of sets since all our sets here will be measurable.

In calculus we saw how to make a change of variables ("*u-du* substitution") in order to simplify certain integrals. There is an analogue of this change of variables in higher-dimensional integration which we now describe. Suppose that B is an $N \times N$ nonsingular matrix over $\mathbb{R}$. If we let $\vec{x} = B\vec{\theta}$, then $d\vec{x} = |\det(B)|d\vec{\theta}$. Thus

$$\int_{\mathbb{R}^N} f(\vec{x})\, d\vec{x} = |\det(B)| \int_{\mathbb{R}^N} f(B\vec{\theta})\, d\vec{\theta}.$$

The coefficient $|\det(B)|$ is referred to as the *Jacobian*. Putting these ideas together, we are able to prove the following result.

THEOREM 13.6. *The volume of a fundamental parallelepiped* $\mathcal{D}$ *is* $d(\Lambda)$. *Thus the volume of a fundamental parallelepiped of a lattice is a well-defined notion.*

We now turn our attention to certain *geometric* sets in $\mathbb{R}^N$. Let $K \subseteq \mathbb{R}^N$. We say that K is *symmetric* if $K = (-1)K$, or alternatively, if $\vec{x} \in K$, then $-\vec{x} \in K$. We say that K is *convex* if whenever $\vec{x}$ and

$\vec{y}$ are in K, then

$$\theta\vec{x} + (1-\theta)\vec{y} \in K \quad \text{for all } \theta, \ 0 \le \theta \le 1.$$

13.7. What is the geometric interpretation of the definition of convexity? Justify your answer.

LEMMA 13.8. *Let K be a nonempty subset of $\mathbb{R}^N$. Then K is both convex and symmetric if and only if for each $\vec{x} \in K$ and $\vec{y} \in K$, it follows that*

$$\theta\vec{x} + \varphi\vec{y} \in K,$$

for all θ and φ such that $|\theta| + |\varphi| \le 1$.

13.9. Show that if $K \subseteq \mathbb{R}^N$ is a nonempty, convex, symmetric set and $\Lambda \subseteq \mathbb{R}^N$ is a lattice. Then $K \cap \Lambda$ is nonempty.

The first result in the geometry of numbers was given by Minkowski in 1896. It gives a remarkably simple test to ensure a convex, symmetric set contains a nonzero lattice point. His result has important consequenes in a surprisingly broad range of mathematics. Although its proof involves several lemmas that follow, we first state Minkowski's Theorem.

MINKOWSKI'S CONVEX BODY THEOREM. *Let $\Lambda \subseteq \mathbb{R}^N$ be a lattice and $K \subseteq \mathbb{R}^N$ a convex symmetric set. If*

$$\mathrm{Vol}_N(K) > 2^N d(\Lambda),$$

then K contains a lattice point $\vec{\lambda} \in \Lambda$ with $\vec{\lambda} \ne \vec{0}$.

We begin our proof with a result due to H. Blichfeldt from 1914.

13.10. BLICHFELDT'S LEMMA. *Let $\mathcal{S} \subseteq \mathbb{R}^N$ be a measurable set. If $\mathrm{Vol}_N(\mathcal{S}) > 1$, then $\mathcal{S}$ contains two distinct points $\vec{s}_1$ and $\vec{s}_2$ such that $\vec{s}_1 - \vec{s}_2 \in \mathbb{Z}^N \setminus \{\vec{0}\}$.*

[$\star$ $\star$ $\star$ $\star$ HINT]

13.11. Give a geometric interpretation of Blichfeldt's Lemma and its proof. Sketch several examples of sets $\mathcal{S}$ and show how the lemma (and proof) applies for the examples.

Module 14

Simultaneous diophantine approximation

The Convex Body Theorem leads to many important results in diophantine approximation. A number of them follow from an extremely useful special case of the Convex Body Theorem, known as Minkowski's Linear Forms Theorem. In this module we will investigate the Linear Forms Theorem and see how it can be applied to produce a geometric proof of Dirichlet's Theorem. We will then discover that it also provides us with the machinery to deduce simultaneous diophantine approximation results that generalize and extend previous theorems.

Suppose we have N different linear forms (homogeneous linear equations) in N variables. Minkowski's Linear Forms Theorem asserts that if this $N \times N$ system of linear forms is nonsingular, then we can always find a nonzero lattice point that simultaneously makes each linear form small (relative to the given data).

In order to prepare us for Minkowski's result, we first consider some relevant background. We begin by examining the supremum norm. We define the *supremum norm*, denoted by $|\ |_\infty$, on $\mathbb{R}^N$ by

$$|\vec{y}|_\infty = \max_{1 \le n \le N} \{|y_n|\}.$$

To illustrate the utility of Minkowski's Linear Forms Theorem, we return to Dirichlet's Theorem (Theorem 3.3) and discover that Dirichlet's result is actually a special case of the Linear Forms Theorem. Verify this remark by using Minkowski's Linear Forms Theorem to give a different (geometric) proof of

14.9. DIRICHLET'S THEOREM (Revisited). *Let α be a real number and $Q \geq 1$ an integer. Then there exists a rational number p/q such that $1 \leq q \leq Q$ and*

$$\left| \alpha - \frac{p}{q} \right| \leq \frac{1}{q(Q+1)}.$$

[$\star\star$ HINT]

We can restate the conclusion of Dirichlet's Theorem as $|\alpha q - p| \leq (Q+1)^{-1}$. That is, p is the nearest integer to αq. This remark inspires the following reformulation.

DIRICHLET'S THEOREM (Reformulated). *Let $L(x)$ be one linear form in one variable given by $L(x) = \alpha x$. Then there exists an integer q satisfying $1 \leq |q| \leq Q$ and $\|L(q)\| \leq (Q+1)^{-1}$.*

Clearly this reformulation is equivalent to the original version of Dirichlet's Theorem. This alternate version, however, allows us to consider more general results that extend Dirichlet's Theorem to higher dimensions.

THEOREM 14.10. *Let $L_1(\vec{x}), L_2(\vec{x}), \ldots, L_M(\vec{x})$ be M real linear forms in N variables defined by*

$$L_m(\vec{x}) = \sum_{n=1}^{N} \alpha_{mn} x_n, \quad m = 1, 2, \ldots, M.$$

Suppose that $Q \geq 1$ is an integer. Then there exists a lattice point $\vec{q} \in \mathbb{Z}^N$ such that $1 \leq |\vec{q}|_\infty \leq Q$ and

$$\max_{1 \leq m \leq M} \{\|L_m(\vec{q})\|\} \leq (Q+1)^{-\frac{N}{M}}.$$

[$\star \star \star$ HINT]

The previous theorem shows that we can find one approximate, $\vec{q} \in \mathbb{Z}^N$, that simultaneously makes each linear form $L_m(\vec{x})$ very close to an integer.

14.11. Show that Theorem 14.10 immediately implies Dirichlet's Theorem.

Suppose we are given L real numbers. Is it possible that there exists *one* nonzero integer q whose product with each of the L real numbers produces quantities that are simultanously all very close to integers? Since we have no information about the L real numbers, upon first inspection it may appear that such a q might not exist. Upon second inspection, however, we can say much more.

14.12. Let $\alpha_1, \alpha_2, \ldots, \alpha_L$ be L real numbers and let $Q \geq 1$ be an integer. Discover a theorem stating that there exists an integer q such that $1 \leq q \leq Q$, and such that

$$\max\{\|\alpha_1 q\|, \|\alpha_2 q\|, \ldots, \|\alpha_L q\|\} \qquad (14.1)$$

is small as a function of Q. Prove your result.

We close our discussion with a notoriously difficult open problem raised by Littlewood. The many attempts by mathematicians to prove Littlewood's conjecture have led to an enormous body of important mathematics. Unfortunately, that body of work does not yet include a proof of the conjecture.

LITTLEWOOD'S CONJECTURE. *Let α and β be real numbers. Then there exists a sequence of positive integers $q_1, q_2, \ldots$ such that*

$$\lim_{n \to \infty} q_n \|\alpha q_n\| \|\beta q_n\| = 0.$$

14.13. Show that the Littlewood Conjecture is true if either α or β is *not* badly approximable.

Thus we need only prove the conjecture for pairs of badly approximable numbers. The set of badly approximable numbers, in fact, has measure zero (meaning that if we were to pick a real number at random, the probability that the number is badly approximable is 0). Therefore in 14.13 you've shown that the conjecture holds for all

LEMMA 15.3. *If $n_1, n_2, \ldots, n_L$ are integers each expressible as the sum of four squares, then their product $n_1 n_2 \cdots n_L$ is a sum of four perfect squares.*

LEMMA 15.4. *Prove that if every prime is the sum of four perfect squares, then every nonnegative integer is the sum of four perfect squares.*

In view of Lemma 15.4, to prove that every positive integer n is expressible as the sum of four perfect squares, we need only consider the case when n is a prime. Clearly the prime 2 is expressible as the sum of four squares: $2 = 1^2 + 1^2 + 0^2 + 0^2$. Thus we need only consider odd primes p.

LEMMA 15.5. *If p is an odd prime, then the congruence*

$$1 + u^2 + v^2 \equiv 0 \bmod p$$

has a solution with $0 \le u < \frac{p}{2}$ and $0 \le v < \frac{p}{2}$.

$[\star\star$ HINT$]$

LEMMA 15.6. *Let p be an odd prime and suppose that u and v are integers such $1 + u^2 + v^2 \equiv 0 \bmod p$. Let T be the matrix defined by*

$$T = \begin{pmatrix} 1 & 0 & 0 & 0 \\ 0 & 1 & 0 & 0 \\ u & v & p & 0 \\ -v & u & 0 & p \end{pmatrix}. \qquad (15.1)$$

If the lattice $\Lambda \subseteq \mathbb{R}^4$ is given by $\Lambda = \left\{ T\vec{x} : \vec{x} \in \mathbb{Z}^4 \right\}$, then $d(\Lambda) = p^2$.

Let $\mathcal{S}(r)$ denote the ball in $\mathbb{R}^4$ of radius r, centered at the origin. That is,

$$\mathcal{S}(r) = \left\{ \vec{x} \in \mathbb{R}^4 : x_1^2 + x_2^2 + x_3^2 + x_4^2 \le r^2 \right\}.$$

LEMMA 15.7. *Let p be an odd prime and let Λ be the lattice in $\mathbb{R}^4$ as defined in Lemma 15.6. Then there exists a lattice point $\vec{\lambda} \in \Lambda$, $\vec{\lambda} \ne \vec{0}$, such that $\vec{\lambda} \in \mathcal{S}\left(\sqrt{1.9p}\right)$.*

$[\star$ HINT$]$

COROLLARY 15.8. *Let p be an odd prime and let u and v be as in Lemma 15.5. Then there exist integers $\lambda_1, \lambda_2, \lambda_3, \lambda_4$ such that*

$$0 < \lambda_1^2 + \lambda_2^2 + (u\lambda_1 + v\lambda_2 + p\lambda_3)^2 + (-v\lambda_1 + u\lambda_2 + p\lambda_4)^2 < 2p.$$

[⋆ HINT]

LEMMA 15.9. *Given the notation of Corollary 15.8, it follows that*

$$\lambda_1^2 + \lambda_2^2 + (u\lambda_1 + v\lambda_2 + p\lambda_3)^2 + (-v\lambda_1 + u\lambda_2 + p\lambda_4)^2 = p. \quad (15.2)$$

[⋆ HINT]

Pulling all the results of this module together, we are able to prove Lagrange's result:

THEOREM 15.10. *Every positive integer is expressible as the sum of four perfect squares.*

Big Picture Question. Once again we see geometry and inequalities come together to produce identities. Give an overview of this theme. Why does it appear that we are "getting out more than we are putting in"?

Further Challenge. If p is a prime congruent to 1 modulo 4, then p is the sum of two perfect squares.

[⋆ ⋆ ⋆ HINT]

16.1. We can view the "fractional part function" as a function on the circle group $\mathbb{R}/\mathbb{Z}$. That is, we can view the interval $[0, 1)$ as a closed circle where the end points 0 and 1 are joined together. For a real number α, describe the sequence $\{\alpha\}, \{2\alpha\}, \{3\alpha\}, \ldots$ in terms of rotating around the circle group $\mathbb{R}/\mathbb{Z}$.

16.2. Suppose that α is a rational number, say $\alpha = p/q$. Is the sequence $\{\alpha\}, \{2\alpha\}, \{3\alpha\}, \ldots$ dense or discrete or neither on the circle? What are all the accumulation points of the sequence? What do you conclude about the fractional parts of integer multiples of rational numbers?

We now consider the more interesting situation when α is irrational. We begin with two useful lemmas.

LEMMA 16.3. *Let α be an irrational number and let $\varepsilon > 0$. Then there exist integers p and q such that*

$$\|\alpha q\| = |\alpha q - p| < \varepsilon.$$

LEMMA 16.4. *Let α be an irrational number and let $\varepsilon > 0$. Then there exists an N such that the sequence*

$$\{\alpha\}, \{2\alpha\}, \{3\alpha\}, \ldots, \{N\alpha\}$$

will partition the circle into intervals each having length less than or equal to ε.

THEOREM 16.5. *A real number α is irrational if and only the sequence*

$$\{\alpha\}, \{2\alpha\}, \{3\alpha\}, \ldots$$

is dense on the circle group $\mathbb{R}/\mathbb{Z}$.

Thus we see another diophantine condition for ensuring irrationality. As a consequence of this result we can deduce the following.

COROLLARY 16.6. *Let α be an irrational real number and $\varepsilon > 0$. Then for any real number β, there exist integers p and q such that*

$$|\alpha q - p - \beta| < \varepsilon.$$

Many of the diophantine approximation results we found in previous modules revolved around making integer linear combinations of α and 1 small (*i.e.*, close to 0). The circle of questions involving integer linear combinations of α and 1 approximating 0 is known as *homogeneous* diophantine approximation. Corollary 16.6 is an example of an *inhomogeneous diophantine approximation* result since it involves an integer linear combination of α and 1 that approximates an *arbitrary* number β (rather than 0). That is, we wish to find integers p and q such that $\alpha q - p$ is approximately β.

A corollary of Dirichlet's Theorem is that for an irrational α there are infinitely many pairs of integers (p, q) such that $|\alpha q - p| < \frac{1}{q}$. We will now explore an inhomogeneous analogue of this result. The inhomogeneous result that follows is due to Kronecker. We begin, as usual, with a useful lemma.

LEMMA 16.7. *Let $\alpha, \beta \in \mathbb{R}$, where α is an irrational number. Then there exist infinitely many pairs of relatively prime integers (m, n), $n > 0$, such that*

$$|\alpha n - m| < \frac{1}{n}.$$

Furthermore, for each pair (m, n), there exists an integer N such that

$$|\beta n - N| \leq \frac{1}{2},$$

and N may be expressed as $N = vm - un$, where u and v are integers with $|v| \leq \frac{1}{2}n$.

[⋆⋆ HINT]

16.8. KRONECKER'S THEOREM. *If α is an irrational real number and β is an arbitrary real number, then there exist infinitely many pairs of integers (p, q), $q > 0$, such that*

$$|\alpha q - p - \beta| < \frac{3}{q}.$$

[⋆⋆ HINT]

We have seen that for an irrational α, the sequence $\{\alpha\}, \{2\alpha\}, \{3\alpha\}, \ldots$ is dense in the unit interval $[0, 1)$. We will now discover

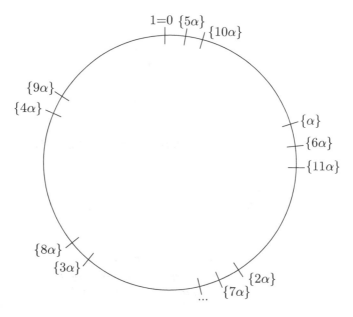

THEOREM 16.15. *Let α be an irrational number. Then the sequence $\{\alpha\}, \{2\alpha\}, \{3\alpha\}, \ldots$ is uniformly distributed modulo 1.*

[⋆ ⋆ ⋆ HINT]

Big Picture Question. What are the differences between a set being dense and a set being uniformly distributed? Discuss the connections among rationality, density, and uniform distribution.

Further Challenge. Suppose we have a square billiard table with side lengths equal to one yard, and not containing any holes (that is, we have bumpers all around the table). A billiard ball is put into motion on this table, bouncing off the sides as it bumps against them. Finally suppose that the table is frictionless so the ball never stops. When will the path of the ball be dense in the square? When will the path not be dense? Can you justify any of your answers?

Module 17

A whole new world of p-adic numbers

As we approach the last few modules of this tract, we turn our attention to the most fundamental concept we've employed throughout our exploration—the notion of *distance*. From the very dawn of our journey, the overt theme was to approximate distances between real and rational numbers. We measured these distances using the absolute value on the real numbers since that is the natural metric (distance measure) on $\mathbb{R}$. This remark raises a basic question: How did the real numbers enter our analysis? We began by studying the rational numbers and their complexity (developing the notion of "height"). The rationals were our footholds into the more mysterious realm of the irrational. But why were we looking at the rational numbers as a subset of the reals? The answer appears clear: because the rationals *are* a subset of the reals! Perhaps, however, the rational numbers are subsets of other interesting, mysterious realms.

In this module we begin a journey into the basic idea of distance and discover new worlds of numbers that are as natural and as important as the reals but have a foreign feel and look. These new numbers, in fact, lead to a broader and deeper understanding both of number and diophantine analysis.

A map $\| \ \| : \mathbb{Q} \to [0, +\infty)$ is called an *absolute value* if it satisfies the following properties:

Inequality (17.1) is called the *strong triangle inequality* and it illustrates the dramatic difference between the usual absolute value and these p-adic absolute values. An absolute value that does *not* satisfy the strong triangle inequality is called *archimedean*. Any absolute value that does satisfy the strong triangle inequality is called *nonarchimedean*.

17.7. Show that the usual absolute value on $\mathbb{Q}$ is archimedean.

As we will discover in the next module, the strong triangle inequality allows an easier and more straightforward resolution of p-adic analytical issues than the analogous archimedean ones. Thus, although nonarchimedean absolute values are strange-looking creatures, they lend themselves beautifully to analysis. As an illustration of the unusual nature of these new absolute values, we'll now see that the geometry of the rationals reveals some surprising structure when considered with respect to the p-adic absolute value. We begin with the following observation regarding when we have equality in the strong triangle inequality.

LEMMA 17.8. *Let p be a prime number. Suppose that $x, y \in \mathbb{Q}$ and $|x|_p \neq |y|_p$. Then*

$$|x + y|_p = \max\{|x|_p, |y|_p\} \ .$$

[$\star\star$ HINT]

17.9. Does the converse of Lemma 17.8 hold in general? Justify your answer.

Let x, y, and z be three rational numbers. If we view these points as vertices of a triangle, we can measure the lengths of its three sides by computing the distance between each pair of points. Now there may be some people who say, "Those points don't form a triangle—they all live on the line!" Well, those people have not yet freed themselves of the shackles of the usual absolute value. They are implicitly viewing the rationals as living inside the set of real numbers and thus have already fixed their measure of distance: the usual archimedean absolute value. If we only consider the rationals and do not fix a distance measure, then we just have a big, unordered collection of fractions, and therefore it is reasonable to consider a triangle having three rational numbers as vertices.

17.10. Consider the rational triangle having vertices $-\frac{1}{2}, \frac{1}{14}$, and $\frac{5}{4}$. Find the lengths of the sides of this triangle when measured 7-adically. What kind of triangle is it? Pick three other rational points and find the lengths of the corresponding triangle 7-adically. Make a conjecture based on your observations.

It is amusing to discover that, in fact, all p-adic triangles are isosceles!

THEOREM 17.11. *Let p be a prime number and x, y, and z be rational numbers. Then the triangle having vertices x, y, and z, when measured p-adically, is isosceles.*

[⋆ HINT]

In fact, as is implied by the next corollary, the two equal side lengths must always be greater than or equal to the third side. Can you see how this remark follows from the corollary below?

COROLLARY 17.12. *If $|x|_p < |y|_p$, then $|x - y|_p = |y|_p$.*

The completion of $\mathbb{Q}$ with respect to the p-adic absolute value, is the set containing $\mathbb{Q}$ in which all p-adic Cauchy sequences from $\mathbb{Q}$ converge. This large set, denoted by $\mathbb{Q}_p$ is called the set of *p-adic numbers*. In the next module we will begin to develop a better feel for the elements of $\mathbb{Q}_p$. The set $\mathbb{Q}_p$ is a nonarchimedean analogue of the set of real numbers.

Given a point $a \in \mathbb{Q}_p$ and a positive real number r, we define the (open) *p-adic ball centered at a of radius r*, $B_p(a, r)$, by

$$B_p(a, r) = \{x \in \mathbb{Q}_p : |x - a|_p < r\} .$$

THEOREM 17.13. *Let p be a prime number, $a \in \mathbb{Q}_p$, and r be a positive real number. If $b \in B_p(a, r)$, then $B_p(a, r) = B_p(b, r)$.*

Module 18

A glimpse into p-adic analysis

In this module we will develop a better sense of the analytical structure of p-adic numbers. Along the way we will discover irrational p-adic numbers, which will inspire us to consider diophantine approximation issues in this new setting. We will illustrate such results at the end of this module.

Recall that the set of p-adic numbers $\mathbb{Q}_p$ is the completion of $\mathbb{Q}$ with respect to the p-adic absolute value $| \ |_p$. Thus it follows that $\mathbb{Q}$ is a dense subset of $\mathbb{Q}_p$.

LEMMA 18.1. *Let $\alpha \in \mathbb{Q}_p$ and ε be any positive real number. Then there exists an $a \in \mathbb{Q}$ such that $|\alpha - a|_p < \varepsilon$.*

What do elements of $\mathbb{Q}_p$ look like? We can represent p-adic numbers as infinite series of a certain form. This representation is analogous to the decimal expansion of the real numbers.

THEOREM 18.2. *Suppose that $a_n \in \mathbb{Q}_p$ for all $n \geq 1$. Then the infinite series $\sum_{n=1}^{\infty} a_n$ converges in $\mathbb{Q}_p$ if and only if $\lim_{n \to \infty} |a_n|_p = 0$.*

[⋆ HINT]

Thus the theory of infinite series is infinitely easier in the nonarchimedean case than the archimedean one. In the archimedean case,

LIOUVILLE'S THEOREM. *Let $\alpha \in \mathbb{Q}_p$ be algebraic of degree $d \geq 2$. Then there exists a positive real constant $c = c(\alpha, p)$ such that for all rationals r/s,*

$$\frac{c}{h(r/s)^d} \leq \left| \alpha - \frac{r}{s} \right|_p .$$

18.11. Show that $\sum_{n=0}^{\infty} p^{n!}$ is a transcendental p-adic number.

Big Picture Question. Recall that $\sum_{n=1}^{\infty} \frac{1}{n}$ diverges to infinity with respect to the usual absolute value even though the individual terms in the series approach zero. We've seen that this phenomenon cannot happen with respect to nonarchimedean absolute values. Where does the nonachimedean argument of this fact break down when adopted for the archimedean setting? Discuss some major differences between archimedean and nonarchimedean analysis.

(See Appendix 1 for some commentary on this question.)

Further Challenge. Construct an infinite series of positive rational numbers that converges in $\mathbb{Q}_p$ for *all* primes p, or prove that such a series cannot exist. If you are amused or interested in this question, take a look at [3].

Module 19

A new twist on Newton's method

To set the stage, let's ask two very basic questions: What does the square root of 2 equal? What does the square root of -1 equal? A naive first response is, "Well, $\sqrt{2} = 1.41421\cdots$, and $\sqrt{-1}$ does not exist." The slightly less naive response, "Clearly, $\sqrt{2} = 1.41421\cdots$ and $\sqrt{-1}$ is an imaginary number", brings a critical issue into focus: what an algebraic number "equals", or whether an algebraic number exists, depends on the field in question. For example, the square root of 2 does not exist in the field of rational numbers, but does exist in the field of real numbers. The square root of -1 does not exist in the field of real numbers, but does exist in the field of complex numbers. In studying algebraic numbers it is valuable to recall a well-known maxim from the real estate world regarding the three most important features of a house: "Location, location, location."

In fact, "the square root of 2" is a bit of a misnomer since in the field of real numbers, there are *two* square roots of 2. Thus it is better to ask, "Is there a solution to $x^2 - 2 = 0$?" Notice that this question can be asked with respect to any field. In particular, is there a solution to $x^2 - 2 = 0$ in $\mathbb{Q}_5$? Is there a solution to $x^2 + 1 = 0$ in $\mathbb{Q}_5$? In this module we consider the algebraic structure of the p-adic numbers and explore conditions for when polynomial equations

possibility disappears. We state Hensel's result here (known as a "lemma"), and then prove it through a sequence of smaller lemmas.

HENSEL'S LEMMA.   *Let $f(x) \in \mathbb{Z}_p[x]$, and suppose there exists an $\alpha_0 \in \mathbb{Z}_p$ such that $|f(\alpha_0)|_p < |f'(\alpha_0)|_p^2$, where f' denotes the formal derivative of f. Then there exists an $\alpha \in \mathbb{Z}_p$ satisfying $f(\alpha) = 0$.*

Note. The "formal derivative", $f'(x)$, is just the function we would produce if we pretended to be first semester calculus students and just used the standard tricks to take the derivative of $f(x)$. So, for example, if $f(x) = 5x^3 - 2x^2 + 3x - 7$, then $f'(x) = 15x^2 - 4x + 3$.

The proof actually contains an algorithm for producing the approximating sequence $\{\alpha_n\}$.

LEMMA 19.7. *Suppose that $f(x) \in \mathbb{Z}_p[x]$ and $\alpha \in \mathbb{Z}_p$. Then $f(\alpha) \in \mathbb{Z}_p$, that is, $|f(\alpha)|_p \leq 1$.*

LEMMA 19.8. *Let $f(x) = \sum_{n=0}^{N} c_n x^n \in \mathbb{Z}_p[x]$. Then for any unknowns X and Y,*

$$f(X + Y) = f_0(X) + f_1(X)Y + f_2(X)Y^2 + \cdots + f_N(X)Y^N \ ,$$

for some polynomials $f_n(X) \in \mathbb{Z}_p[X]$. Moreover, $f_0(X) = f(X)$ and $f_1(X) = f'(X)$, where $f'(X)$ denotes the formal derivative of the polynomial $f(X)$.

LEMMA 19.9. *Let $f(x) \in \mathbb{Z}_p[x]$ and suppose there exists an $\alpha_0 \in \mathbb{Z}_p$ such that $|f(\alpha_0)|_p < |f'(\alpha_0)|_p^2$. Then $-f(\alpha_0)/f'(\alpha_0) \in \mathbb{Z}_p$.*

LEMMA 19.10. *Let $f(x) \in \mathbb{Z}_p[x]$ and suppose there exists an $\alpha_0 \in \mathbb{Z}_p$ such that $|f(\alpha_0)|_p < |f'(\alpha_0)|_p^2$. Let $\beta_0 = -f(\alpha_0)/f'(\alpha_0)$. Then*

$$|f(\alpha_0 + \beta_0)|_p \leq |f(\alpha_0)|_p \ .$$

[⋆ HINT]

LEMMA 19.11. *Given the notation in Lemma 19.10, it follows that*

$$|f'(\alpha_0 + \beta_0) - f'(\alpha_0)|_p < |f'(\alpha_0)|_p \ .$$

[⋆⋆ HINT]

LEMMA 19.12. *Given the notation in Lemma 19.10, it follows that*

$$|f'(\alpha_0 + \beta_0)|_p = |f'(\alpha_0)|_p .$$

[⋆ HINT]

LEMMA 19.13. *Given the notation in Lemma 19.10, there exist two sequences in $\mathbb{Z}_p$, $\alpha_0, \alpha_1, \alpha_2, \ldots$ and $\beta_0, \beta_1, \beta_2, \ldots$, such that $\beta_n = -f(\alpha_n)/f'(\alpha_n)$, $\alpha_{n+1} = \alpha_n + \beta_n$, and*

$$|f(\alpha_{n+1})|_p < |f'(\alpha_{n+1})|_p^2 \text{ for all } n \geq 0.$$

Moreover, $|f(\alpha_{n+1})|_p < |f(\alpha_n)|_p$ for all $n \geq 0$.

[⋆ ⋆ ⋆ HINT]

From the previous lemma, we notice that

$$\alpha_{n+1} = \alpha_n - \frac{f(\alpha_n)}{f'(\alpha_n)} ,$$

which resembles the sequence generated by Newton's method. We are now ready to prove the p-adic analogue.

19.14. HENSEL'S LEMMA. *Let $f(x) \in \mathbb{Z}_p[x]$, and suppose there exists an $\alpha_0 \in \mathbb{Z}_p$ such that $|f(\alpha_0)|_p < |f'(\alpha_0)|_p^2$, where f' denotes the formal derivative of f. Then there exists an $\alpha \in \mathbb{Z}_p$ satisfying $f(\alpha) = 0$.*

[⋆⋆ HINT]

Notice that the word "usually" does not appear in Hensel's Lemma as it did in Newton's method. In the p-adic case, if we can approximate a zero, then we definitely will converge to it!

Big Picture Question. Suppose that p is a prime, $p \neq 2$, and $b \in \mathbb{Z}_p$ satisfies $|b|_p = 1$. Assume that there exists an $\alpha_0 \in \mathbb{Z}_p$ such that $|\alpha_0^2 - b|_p < 1$. Then prove that there exists an element $\alpha \in \mathbb{Z}_p$ satisfying $\alpha^2 = b$. Use your result to determine if $x^2 + 1$ has a solution in $\mathbb{Q}_5$.

(See Appendix 1 for some commentary on this question.)

completion of $\mathbb{Q}$ with respect to that absolute value, then we produce either $\mathbb{R}$ or $\mathbb{Q}_p$, for some prime p. These complete fields are called *local fields* since they are the extension of $\mathbb{Q}$ for which the Cauchy sequences for a *particular* absolute value all converge (in some sense, we are "localizing" $\mathbb{Q}$ by introducing a distance function). Thus, our simple argument above is an example where analysis within a local field gives rise to information about the global field. This theme is known as a "local-to-global" technique and was first observed by Hasse in 1931. In this final module, we describe this type of a result and illustrate its utility with an example.

The "local-to-global" method is particularly well suited for analyzing quadratic forms. A *quadratic form* is a polynomial $f(\vec{x}) \in \mathbb{Q}[\vec{x}]$ in N variables, where

$$\vec{x} = \begin{pmatrix} x_1 \\ x_2 \\ \vdots \\ x_N \end{pmatrix},$$

having the general shape

$$f(\vec{x}) = \sum_{i=1}^{N} \sum_{j=1}^{N} a_{ij} x_i x_j, \quad \text{where } a_{ij} \in \mathbb{Q}.$$

For example, $f(x,y) = x^2 - 10xy + y^2$, $f(x,y,z) = x^2 + y^2 + z^2$, and $f(x,y,z,w) = 5x^2 - 9xy - 7y^2 + 2yz - 4yw - 5zw + 4w^2$ are all examples of quadratic forms, while $g(x,y) = x^3 + y^3$ and $g(x,y,z) = x^2 + xyz - z^2$ are not. If $f(\vec{x})$ is a quadratic form, then clearly $\vec{x} = \vec{0}$ is always a solution to $f(\vec{x}) = 0$. We call this solution the *trivial solution*.

Suppose that $f(\vec{x}) \in \mathbb{Q}[\vec{x}]$ is a quadratic form. Then the Hasse-Minkowski Theorem asserts that if we can find a local, nontrivial solution to $f(\vec{x}) = 0$ in each local field over $\mathbb{Q}$, then there must exist a nontrivial solution in $\mathbb{Q}^N$. The proof of this theorem is extremely difficult and we will not consider the argument here. Instead, we'll see the result in action.

HASSE-MINKOWSKI THEOREM.   *Let $f(\vec{x}) \in \mathbb{Q}[\vec{x}]$ be a quadratic form in N variables. There exists a nontrivial solution $\vec{x} \in \mathbb{Q}^N$ to*

$f(\vec{x}) = 0$ if and only if there exists an $\vec{x}_0 \in \mathbb{R}^N \setminus \left\{\vec{0}\right\}$ such that $f(\vec{x}_0) = 0$, and for each prime p, there exists an $\vec{x}_p \in \mathbb{Q}_p^N \setminus \left\{\vec{0}\right\}$ such that $f(\vec{x}_p) = 0$.

In other words, a local solution over every completion of $\mathbb{Q}$, implies a global solution in $\mathbb{Q}^N$. The theorem also asserts that the converse is also true, but this fact is very easy, as we now verify for ourselves.

20.1. Given the quadratic form $f(\vec{x}) \in \mathbb{Q}[\vec{x}]$, suppose that there is a nonzero vector $\vec{x} \in \mathbb{Q}^N$ satisfying $f(\vec{x}) = 0$. Then show that there must exist a nonzero vector $\vec{x}_0 \in \mathbb{R}^N$ such that $f(\vec{x}_0) = 0$, and for each prime p, a nonzero vector $\vec{x}_p \in \mathbb{Q}_p^N$ such that $f(\vec{x}_p) = 0$.

COROLLARY 20.2. Let $f(\vec{x}) \in \mathbb{Q}[\vec{x}]$ be a quadratic form in N variables, and let $a \in \mathbb{Q}$, $a \neq 0$. There exists a nontrivial solution $\vec{x} \in \mathbb{Q}^N$ to $f(\vec{x}) = a$ if and only if there exists a nontrivial solution in $\mathbb{R}^N$ and a nontrivial solution in $\mathbb{Q}_p^N$ for each prime p.

[⋆ HINT]

We now illustrate the utility of the Hasse-Minkowski Theorem by considering a slight generalization of the diophantine equation we considered at the opening of this module. Specifically, we now wish to find all the integers m such that the diophantine equation

$$x^2 + y^2 = mz^2$$

has nontrivial solutions x, y, z in $\mathbb{Q}$. Equivalently (but geometrically), we wish to find all integers m such that the circle centered at the origin of radius $\sqrt{m}$ contains rational points. We begin by showing that we need only consider the case when m is a square-free integer.

LEMMA 20.3. Suppose that $m = m't^2$, where m, m', and t are positive integers. Then there exists a nontrivial integer solution (x, y, z) to $x^2 + y^2 = mz^2$ if and only if there exists a nontrivial integer solution (x, y, z) to $x^2 + y^2 = m'z^2$.

In view of Lemma 20.3, we can assume that m is a square-free integer. That is, m is a product of distinct primes. We will adopt the

$m \equiv 3 \bmod 4$. *If m is even, then there exist $x, y, z \in \mathbb{Q}_2$, not all zero, satisfying $x^2 + y^2 = mz^2$ if and only if $m \equiv 6 \bmod 8$.*

We can now summarize our findings and produce necessary and sufficient conditions on m so that there are nonzero, rational solutions to the diophantine equation $x^2 + y^2 = mz^2$.

THEOREM 20.12. *Let m be a fixed integer. There exist $x, y, z \in \mathbb{Q}$, not all zero, satisfying*
$$x^2 + y^2 = mz^2$$
if and only if the following conditions hold:
  (i) m is a positive integer.
 (ii) either $m \equiv 3 \bmod 4$ or $m \equiv 6 \bmod 8$.
(iii) the only primes p dividing m satisfy $p \equiv 1 \bmod 4$.

Thus we see that the slogan of some environmentally minded people, "Think globally, act locally", is an important philosophy within the world of diophantine analysis.

Big Picture Question. In Module 17 we described Ostrowski's result stating that the usual archimedean absolute value and the p-adic absolute values are the only nontrivial absolute values on $\mathbb{Q}$. In the Big Picture Question from Module 17 you were asked to show how all these absolute values work together to yield the "product formula". How does the Hasse-Minkowski Theorem fit into this general theme of cooperation among all the absolute values of $\mathbb{Q}$?

Further Challenge. Let $f(\vec{x})$ be a polynomial in N variables having rational coefficients. Suppose there exists a vector $\vec{q} \in \mathbb{Q}^N$, $\vec{q} \neq \vec{0}$, satisfying $|f(\vec{q})| < 1$ and for each prime p, $|f(\vec{q})|_p \leq 1$. Then prove there exists a vector $\vec{t} \in \mathbb{Z}^N$, $\vec{t} \neq \vec{0}$, such that $f(\vec{t}) = 0$.

Appendix 1

Selected Big Picture Question commentaries

MODULE 1. Given the observations made in Module 1, it seems reasonable to measure the complexity of a rational number (written in lowest terms) by the size of its denominator. Another potential guess is to consider the rational number in its decimal expansion and measure its complexity by the length of its periodic string. One problem with this guess is that the complexity would then depend on which base we used to expand the rational number. We desire a measure that depends only on the intrinsic nature of the rational and not on its representation.

Technically, the measure of complexity of the rational number p/q is known as the *height* of p/q, denoted as $h(p/q)$ and is usually defined by $h(p/q) = \max\{|p|, |q|\}$. Why is the numerator included? The answer is that we would like our measure of complexity to have the property that for any fixed value c, there are only *finitely* many rational numbers satisfying $h(p/q) \leq c$. So for example, there are only 39 rationals that satisfy $h(p/q) \leq 5$. On the other hand, there are infinitely many rationals having their denominators bounded above by 5. However if we further insist that our rational numbers having bounded denominators are to be close to a *fixed* number α, then we would be left with only a finite list. For example, there are only finitely many rationals p/q having denominators bounded above by 5

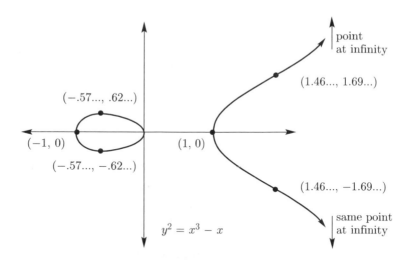

$(1.46\ldots, \pm 1.69\ldots)$. Putting these observations together we get the following graph.

We notice that the upper right portion of the curve appears to be approaching a heading of due north (that is, $\frac{dy}{dx} \to +\infty$, so the tangents are becoming vertical), while the lower right part of the curve seems to be approaching a heading of due south. Thus in the projective completion of $\mathbb{R}^2$, these northern and southern wings will meet at the same point at infinity—the point where all vertical lines meet. Therefore in the projective completion, the curve becomes two closed loops. Given this observation, it would appear that we are looking at a cross-sectional slice (or level curve) of a one-holed torus. Thus a reasonable guess would be that the surface generated by $y^2 = x^3 - x$ is a one-holed torus.

MODULE 14. There is no known continued fractionesque algorithm for finding the complete list of best approximations to simultaneous systems. Recently in [2], an algorithm, similar in spirit to the continued fraction algorithm, has been found to generate the complete list of best approximations for simultaneous systems where all the values to be approximated are from the vector space $\mathbb{Q} + \mathbb{Q}\alpha$ for some $\alpha \in \mathbb{R}$. The best possible constants in the upper bounds (analogous to the Markoff constant in the 1-dimensional case) are also given in that work. The general case remains open.

MODULE 18. If we let $S_N = \sum_{n=1}^{N} \frac{1}{n}$ denote the Nth partial sum of the harmonic series, then for arbitrary $N > M$, $S_N - S_M = \frac{1}{M+1} + \frac{1}{M+2} + \cdots + \frac{1}{N}$. But how large can this quantity be when measured with respect to the usual (archimedean) absolute value? The answer is that it can get very large (on the order of $\log N - \log(M+1)$) and if N is allowed to grow, we cannot keep the difference under any fixed $\varepsilon > 0$. If we try to crash through with the triangle inequality, we would see

$$|S_N - S_M| = \left| \frac{1}{M+1} + \frac{1}{M+2} + \cdots + \frac{1}{N} \right|$$
$$\leq \left| \frac{1}{M+1} \right| + \left| \frac{1}{M+2} \right| + \cdots + \left| \frac{1}{N} \right|.$$

Actually, in this particular case, we have equality in the above inequality. Even though each term on the right-hand side is small, when we add them up, the sum will continue to increase without bound.

However, in the nonarchimedean case we are able to exploit the *strong* triangle inequality and bound the difference of partial sums by the *maximum* value of the individual terms (rather than the sum of all of them). Thus, if those individual terms are approaching zero p-adically, we see that the sequence of partial sums forms a p-adic Cauchy sequence and thus the associated infinite series converges (in $\mathbb{Q}_p$).

MODULE 19. Let $f(x) = x^2 - b$. Thus for the given $\alpha_0 \in \mathbb{Z}_p$, we have that $|f(\alpha_0)|_p < 1$. We know that $|\alpha_0|_p \leq 1$ (since α_0 is in $\mathbb{Z}_p$). We now claim that $|\alpha_0|_p = 1$. To prove this assertion, assume that $|\alpha_0|_p < 1$ (and therefore $|\alpha_0^2|_p < 1$). Since $|b|_p = 1$, Lemma 17.8 implies that $|\alpha_0^2 - b|_p = \max\{|\alpha_0^2|_p, |b|_p\} = |b|_p = 1$. Thus we see that $|f(\alpha_0)|_p = 1$, which contradicts the fact that $|f(\alpha_0)|_p < 1$. Therefore it must be the case that $|\alpha_0|_p = 1$. We now see that $|f'(\alpha_0)|_p = |2\alpha_0|_p = 1$ (since $p \neq 2$). Hence we have that $|f(\alpha_0)|_p < |f'(\alpha_0)|_p^2$ for some p-adic integer α_0. Hensel's Lemma now implies that there must exist a p-adic integer α such that $f(\alpha) = 0$ and this completes our argument.

3.9. Clear denominators of both of the inequalities of the lemma and show that if they are added together one has $4y^2 - 4(\sqrt{5} - 1)xy + (5 - 2\sqrt{5} + 1)x^2 \le 0$. Now factor and deduce a contradiction.

3.10. First show that it is enough to prove the result in the case when $0 < \alpha < 1$. Show that there must exist two adjacent elements of $\mathcal{F}_n$, say a/b and c/d, such that $\frac{a}{b} < \alpha < \frac{c}{d}$. The goal now is to show that one of the fractions $a/b, c/d$, or $(a + c)/(b + d)$ will satisfy the inequality of the theorem. To verify this claim, assume that none of the three fractions satisfy the inequality. Consider two cases. First, when $\alpha < (a + c)/(b + d)$, write down the three inequalities you assume hold and carefully remove the absolute values by taking note of the ordering of the fractions and α. Add these inequalities together to produce two inequalities that resemble the inequalities in Lemma 3.9. This deduction would lead to a contradiction. Adopt a similar strategy for the case when $\alpha > (a + c)/(b + d)$. Now show that we must have infinitely many distinct rationals that satisfy the inequality from the theorem.

3.11(d). Use the triangle inequality.

3.12. Let $\varphi = \frac{1+\sqrt{5}}{2}$ and suppose that for this φ, there exists a real positive number m such that

$$\left| \varphi - \frac{a_i}{b_i} \right| < \frac{1}{mb_i^2}$$

holds for infinitely many distinct a_i/b_i. Show that $b_i \to \infty$ as $i \to \infty$. Now for any a_i/b_i use your results from 3.11 to give upper and lower bounds for $\left| \varphi - \frac{a_i}{b_i} \right| \left| \overline{\varphi} - \frac{a_i}{b_i} \right|$. Show that $m \le \sqrt{5}$.

4.10(b). For the second part of (b), first consider $p_n q_{n-1} - p_{n-1} q_n$.

5.3. Consider identity (4.1) and use Lemma 4.1.

5.4. Again, consider Lemma 4.1.

5.9. To prove part (ii), use Lemma 5.8 to show there are integers x, y that satisfy identity (5.3). Now show that given the hypotheses, $y \ne 0$. Next, consider the case when $x = 0$ and show that (ii) holds.

Finally, consider the case when $xy \neq 0$ and show that it must follow that $xy < 0$. Rewrite the expression $|\alpha q - p|$ using (5.3) and show that there must be equality in the triangle inequality.

5.11. Consider Lemma 5.10 and Corollary 4.5.

5.12. Find nontrival upper and lower bounds for the quantity $|p_{N-1}q - q_{N-1}p|$.

6.2. Consider Lemma 5.4.

6.3. Note that to compute $\mu(\alpha)$, we must consider the limit inferior over *all* positive integers q, not just the denominators of α's convergents. To get around this point, consider Hurwitz's Theorem 3.10 and Legendre's Theorem 5.12.

6.4. Consider the partial quotients of α. If $a_n \geq 3$ for infinitely many n, then examine the limit inferior in Lemma 6.3 for those N for which $a_{N+1} \geq 3$. Show that for such α's, $\mu(\alpha) \leq \frac{1}{3} < \frac{1}{\sqrt{5}}$. Thus we now need only consider α's that are equivalent to irrationals having only 1's and 2's as partial quotients. Suppose that the string "$2, 2$" occurs infinitely often in the continued fraction expansion. Now show that for such α's, $\mu(\alpha) \leq \frac{2}{5} < \frac{1}{\sqrt{5}}$. Finally, show that there is an upper bound for $\mu(\alpha)$ less than $\frac{1}{\sqrt{5}}$ in the case when the string "$1, 2, 1$" occurs infinitely often in the continued fraction expansion for α.

6.5. Refine the method of proof used in Theorem 6.4.

6.10. Consider the neighbors of (k, l, m) and examine its neighbor that has a smaller maximum element. Now repeat the process.

7.6. Start with the case of a purely periodic continued fraction and use Lemma 5.3 or Lemma 5.4 to deduce that α satisfies a quadratic equation with integer coefficients. Use this idea and Lemma 5.3 or Lemma 5.4 (again) to deduce the general case.

7.11. First use Lemma 7.10 to show that m exists.

13.10. *Some steps of the proof—fill in all the details and finish.*
Assume that the sets $\left\{ \mathcal{S} - \vec{\xi} : \vec{\xi} \in \mathbb{Z}^N \right\}$ are disjoint. Let

$$\mathcal{D} = \left\{ \vec{\theta} \in \mathbb{R}^N : 0 \leq \theta_n < 1 \text{ for all } n = 1, 2, \ldots, N \right\}.$$

Show that the sets $\left\{ \mathcal{D} + \vec{\xi} : \vec{\xi} \in \mathbb{Z}^N \right\}$ *are* disjoint and

$$\bigcup_{\vec{\xi} \in \mathbb{Z}^N} \left(\mathcal{D} + \vec{\xi} \right) = \mathbb{R}^N.$$

Verify that

$$\text{Vol}_N(\mathcal{S}) = \text{Vol}_N \left(\mathbb{R}^N \cap \mathcal{S} \right)$$

$$= \text{Vol}_N \left(\left\{ \bigcup_{\vec{\xi} \in \mathbb{Z}^N} \left(\mathcal{D} + \vec{\xi} \right) \right\} \cap \mathcal{S} \right)$$

$$= \text{Vol}_N \left(\bigcup_{\vec{\xi} \in \mathbb{Z}^N} \left\{ \left(\mathcal{D} + \vec{\xi} \right) \cap \mathcal{S} \right\} \right)$$

$$= \sum_{\vec{\xi} \in \mathbb{Z}^N} \text{Vol}_N \left\{ \left(\mathcal{D} + \vec{\xi} \right) \cap \mathcal{S} \right\}$$

$$= \sum_{\vec{\xi} \in \mathbb{Z}^N} \text{Vol}_N \left\{ \left(\mathcal{D} \cap \left(\mathcal{S} - \vec{\xi} \right) \right) + \vec{\xi} \right\}$$

$$= \sum_{\vec{\xi} \in \mathbb{Z}^N} \text{Vol}_N \left\{ \mathcal{D} \cap \left(\mathcal{S} - \vec{\xi} \right) \right\}$$

$$= \text{Vol}_N \left\{ \bigcup_{\vec{\xi} \in \mathbb{Z}^N} \left\{ \mathcal{D} \cap \left(\mathcal{S} - \vec{\xi} \right) \right\} \right\}$$

$$= \text{Vol}_N \left\{ \mathcal{D} \cap \left\{ \bigcup_{\vec{\xi} \in \mathbb{Z}^N} \left(\mathcal{S} - \vec{\xi} \right) \right\} \right\}$$

$$\leq \text{Vol}_N(\mathcal{D}) = 1.$$

Now finish the proof on your own.

13.12. First consider the set $S = \frac{1}{2}L$. Then recall how the volume of a set changes when there is a scaling factor.

13.13. Suppose that $\Lambda = B\mathbb{Z}^N$. Now analyze $L = B^{-1}K$.

13. Further Challenge. Consider $K = \left(1 + \frac{1}{n}\right)\mathcal{K}$, where n is a positive integer. Now apply the Convex Body Theorem to K, let $n \to \infty$, and use the fact that $\mathbb{Z}^N$ is a discrete set.

14.7. Consider the set $K = \{\vec{x} \in \mathbb{R}^N : |E^{-1}C\vec{x}|_\infty \leq 1\}$ and apply Minkowski's Convex Body Theorem.

14.9. Let $C = \begin{pmatrix} -1 & \alpha \\ 0 & 1 \end{pmatrix}$ and $\varepsilon_1 = \frac{1}{Q+\theta}$, where $0 < \theta < 1$. The rest is up to you.

14.10. Let $A = (\alpha_{mn})$ be the $M \times N$ real matrix associated with the system of linear forms and C be the $(M+N) \times (M+N)$ real matrix defined by

$$C = \begin{pmatrix} \mathbf{1}_M & | & -A \\ -- & | & -- \\ \bigcirc & | & \mathbf{1}_N \end{pmatrix},$$

where $\mathbf{1}_M$ denotes the $M \times M$ identity matrix. Set $\varepsilon_1 = \varepsilon_2 = \cdots = \varepsilon_M = (Q + \theta)^{-N/M}$ and $\varepsilon_{M+1} = \varepsilon_{M+2} = \cdots = \varepsilon_{M+N} = (Q + \theta)$, where θ is a real number, $0 < \theta < 1$. Now apply the Linear Forms Theorem with $\Lambda = \mathbb{Z}^{M+N}$.

15.5. Show that the elements of the set $S_1 = \left\{0^2, 1^2, 2^2, \ldots, \left(\frac{p-1}{2}\right)^2\right\}$ are all distinct modulo p. Produce a similar argument for the set

$$S_2 = \left\{-1 - 0^2, -1 - 1^2, -1 - 2^2, \ldots, -1 - \left(\frac{p-1}{2}\right)^2\right\}.$$

Now use the Pigeonhole Principle (Theorem 3.1) to conclude that there must exist an element from S_1 that is congruent modulo p to an element of S_2.

15.7. Note that $\text{Vol}_4(S(r)) = \pi^2 r^4/2$. Now apply Minkowski's Convex Body Theorem.

some $a = 1, 2, 3$, or 4. Check each possible value of a and deduce a contradiction.

19.10. Use Lemma 19.8 and the strong triangle inequality.

19.11. Apply Lemma 19.8 to the polynomial $f'(x)$ and recall the definition of β_0.

19.12. Use Lemma 19.11 and Corollary 17.12.

19.13. First show that the inequalities hold for $n = 0$, and then apply the previous lemmas with α_1 to deduce the result for α_2, and then repeat. It may also be useful to show that $|f'(\alpha_n)|_p = |f'(\alpha_0)|_p$ for all n. Finally note that we can write $|f(\alpha_n)|_p = p^{e_n}$ for some $e_n \in \mathbb{Z}$.

19.14. Show that the sequence α_n from Lemma 19.13 converges. Then show that $|f(\alpha_n)|_p \to 0$ as $n \to \infty$.

20.2. Apply the Hasse-Minkowski Theorem to the quadratic form $g(\vec{x}, z) = f(\vec{x}) - az^2$.

20.5. Let $z_0 = 1$ and adopt a modification of the proof of Lemma 15.5.

20.6. Use Lemma 20.5 to show that $\left| x_0^2 + y_0^2 - mz_0^2 \right|_p \le \frac{1}{p}$. Now consider the quadratic polynomial $f(X) = X^2 + y_0^2 - mz_0^2$ and apply Hensel's Lemma.

20.7. There is a solution with $z = 0$.

20.8. For the first part of the lemma, recall that m is a *square-free* integer. For the second part, apply the first part repeatedly to find the desired $x, y, z \in \mathbb{Z}_p$.

20.9. The hint given for 19.5 might be helpful.

Appendix 3

Further reading

- **Modules 2 & 3:**   [4], [12], [13], [15]
- **Module 4:**   [4], [10], [12], [14], [18], [19]
- **Module 5:**   [1], [12], [15], [18], [19]
- **Module 6:**   [7], [12]
- **Module 7:**   [12], [14], [15], [18], [19]
- **Module 8:**   [12], [14], [18], [19]
- **Module 9:**   [9], [13], [15], [18]
- **Module 10:**   [4], [15]
- **Modules 11 & 12:**   [13], [17]
- **Module 13:**   [5]
- **Module 14:**   [2], [4], [15]
- **Module 15:**   [13]
- **Module 16:**   [4], [9]
- **Modules 17 & 18:**   [3], [6], [8], [11]
- **Module 19:**   [4], [6], [11]
- **Module 20:**   [6], [16]

References

[1]   E.B. Burger, *Diophantine olympics and world champions: Polynomials and primes down under*, Amer. Math. Monthly **107** (2000), Nov.

[2]   E.B. Burger, *On simultaneous diophantine approximation in the vector space* $\mathbb{Q} + \mathbb{Q}\alpha$, J. Number Theory **82** (2000), 12–24.

stereogram cover art. I also thank Robert E. Tubbs, from The University of Colorado at Boulder, who made a number of valuable comments, and Michael Starbird, from The University of Texas at Austin, who encouraged me to pursue this project.

Finally, I wish to express my deepest gratitude to my teacher and mentor Jeffrey D. Vaaler from The University of Texas at Austin. Jeff was my Ph.D. advisor and was the first to introduce me to this area of number theory. He brought these ideas to life and allowed me to discover their richness and beauty. Much of my point of view and approach to this subject is largely due to him. It is a pleasure and honor for me to be able to acknowledge and thank Jeffrey Vaaler.

Index

Selected Titles in This Series